CIVIL ENGINEERING SOCIETAL
&
GLOBAL IMPACT

AICTE'S LIST OF PRESCRIBED TEXTBOOKS

As Per Model Curriculum & Outcome Based Education
aligned with
NATIONAL EDUCATION POLICY 2020

UNDERGRADUATE COURSE BOOKS

S.No.	ISBN	Book Name	Author
1	9789355382160	Advanced Electrical Machines: Theory and Practicals	Bharat Singh Rajpurohit
2	9789355382139	Analog Electronic Circuits: Theory and Practical	Venu Sonti, Sachin Jain
3	9789355383808	Analog Electronic Devices: Theory and Practicals	N. B Balamurugan
4	9789391505158	Basic Electrical Engineering (with Lab Manual)	S.K. Sahdev
5	9789355388087	Basic of Electronics	Brijesh Iyer
6	9789355387127	Basics of Electrical Machines: Theory and Practicals	D. K. Palwalia, U. K. Kalla, R. K. Kumawat
7	9789355389381	Basics of Mechanics	Shubhankar Bhowmick, N. Venkata Swamy Naidu
8	9789355386250	Basics of Thermodynamics	Parmod Kumar, Atul Dhar
9	9789355386984	Biology for Engineers	Annamma Odaneth
10	9789391505141	Chemistry-I	Manisha Agarwal
11	9789355384843	Civil Engineering Societal and Global Impact	Shakuntala Acharya
12	9789355382184	Computer Aided Civil Engineering Drawing	D. Chandra Mouli, B. Swathi
13	9789355382580	Computer Aided Design & Analysis	Hari Kumar Voruganti
14	9789355387592	Computer Architecture	D. Vaithiyanathan
15	9789355385093	Computer Networks	Hari Prabhat Gupta
16	9789355385307	Construction Engineering & Management	Sparsh Johari
17	9789355388728	Control Systems	S.N. Deepa
18	9789355385420	Data Structure and Algorithms	Ranbir Singh Sanasam
19	9789355386700	Design and Analysis of Algorithms	Hari P. Gupta, Rahul Mishra
20	9789355388872	Digital Electronics and System	Abhishek Bhatt
21	9789355386571	Digital Signal Processing	Sanjeev Sharma
22	9789355383396	Disaster management	Pradeep K. Goyal, Anil K. Gupta
23	9789355385222	Discrete Mathematical Structures	Narendra S. Chaudhari, Sharmila S. P.
24	9789355387868	Effective Technical Communication	Pritee Sharma, Sanchita Verma
25	9789355383389	Electrical Circuit Analysis and Network Theory	Shouri Chatterjee
26	9789355388957	Electromagnetic Waves	Uttam M. Pal, Srijith K.
27	9789355384652	Electronic Communication	Vimal Bhatia
28	9789355388711	Embedded Systems	Santanu Chattopadhyay
29	9789355385147	Energy Science and Engineering	M. Rizwan, Majid Jamil
30.	9789355387929	Engineering Economics, Estimation & Costing	Yogesh C. Joshi
31	9789355384669	Engineering Geology	R. N. Khare, Ketan Chourasia
32	9789391505066	Engineering Graphics & Design	Pradeep Jain
33	9789391505097	English (with Lab Manual)	Kulbhushan Kumar
34	9789355385963	Environmental Engineering	Chandra Sekhar Matli
35	9789355388117	Fluid Mechanics	Shreenivas N. Londhe, Pradnya R. Dixit
36	9789355388377	Fluid Mechanics & Fluid Machines	R.P. Saini
37	9789355387103	Geotechnical Engineering	Paramita Bhattacharya

UNDERGRADUATE COURSE BOOKS

S.No.	ISBN	Book Name	Author
38	9789355385130	Hydraulic Engineering	Sayantan Ganguly
39	9789355385550	Hydrology & Water Resources Engineering	Mitthan Lal Kansal, P.K. Agarwal
40	9789355389688	Information Technology Workshop in MATLAB	Hari Prabhat Gupta
41	9789355383235	Instrumentation for Civil Engineering Applications	Ashim Kanti Dey, Manita Das
42	9789355387042	Introduction to Civil Engineering	Manabendra Saharia, Nagendra R. Velaga
43	9789355384805	Introduction to Database Systems	Vikram Singh
44	9789355387509	Machine Element & System Design	Sabah Khan
45	9789355385383	Machine Learning	Himanshu Mittal
46	9789355388414	Manufacturing Automation	Chitragupt Swaroop Chitransh
47	9789355387233	Manufacturing Processes	Dheerendra Kumar Dwivedi
48	9789355389886	Materials Engineering	Dheerendra Kumar Dwivedi
49	9789391505172	Mathematics - I Calculus and Linear Algebra [For Non-Computer Science Engineering Branches]	Reena Garg
50	9789391505271	Mathematics - I Calculus and Linear Algebra [For Computer Science Engineering Branches]	Reena Garg
51	9789355384607	Measurements & Metrology	V.P. Chandramohan
52	9789355383730	Mechanics of Materials	Arvind Vyavahare
53	9789355383358	Mechatronics, Robotics & Control	J. Srinivas
54	9789355385116	Microprocessor and Microcontroller	Saurabh Chaudhury, Risha Mal
55	9789355386083	Microprocessors	Yogendra Kumar Gupta
56	9789355389169	Mobile Communication and Networks	Vimal Bhatia
57	9789355385475	Operating Systems	Sukomal Pal
58	9789355385604	Operating Systems	Sanjaya Kumar Panda
59	9789391505134	Physics (Oscillations, Waves and Optics) With lab manual	A.B. Bhattacharya, Atanu Nag
60	9789391505257	Physics (Quantum Mechanics for Engineers) With lab manual	A.B. Bhattacharya, Atanu Nag
61	9789391505165	Physics (Introduction to Electromagnetic Theory) (with Lab Manual)	A.B. Bhattacharya, Atanu Nag
62	9789391505059	Physics (Introduction to Mechanics) (with Lab Manual)	A.B. Bhattacharya, Atanu Nag
63	9789355385956	Power Electronics: Theory and Practicals	Lalit Chandra Saikia
64	9789355384942	Power Systems-I	Suresh Mikkili
65	9789355386236	Power Systems-II	Arun Rahul S, Manas K. Jena
66	9789355383754	Product Innovation & Entrepreneurship	Vigneswaran C
67	9789355384584	Production & Operation Management	Navin Kumar Dev
68	9789391505219	Programming for Problem Solving (with Lab Manual)	R.S. Salaria
69	9789355385451	Signals and Systems	Sanjay L. Nalbalwar
70	9789355386021	Strength of Materials	Uday Shanker Dixit, Nelson Muthu, S. M. Kamal
71	9789355383792	Structural Engineering	T.D. Gunneswara Rao, D. Ravi Prasad
72	9789355387004	Surveying and Geomatics	P. K. Garg
73	9789355384317	Testing & Evaluation of Civil Engineering Materials	Sandeep Chaudhary, Kashyap A Patel
74	9789355382306	Theory of Computation	Debi Prasanna Acharjya
75	9789355384928	Transportation Engineering	A.K. Raji
76	9789355385208	VLSI Design	Mahendra Sakare, Shonal Chouksey
77	9789391505332	Workshop / Manufacturing Practices (with Lab Manual)	Veeranna D.K.

DIPLOMA COURSE BOOKS

S.No.	ISBN	Book Name	Author
1	9789355382481	Aanandam An Act of Community Service	Shuchi Sharma
2	9789355385567	Advanced Manufacturing Processes	Ranjeet Kumar Sahu, Devendra Laxman Kamble
3	9789355385109	Advanced Surveying: Theory & Practice	Ramakant Agrawal, Parshottam Sarathe
4	9789355384546	Algorithms	Piyoosh P, Arnab Sarkar
5	9789391505448	Applied Chemistry (with Lab Manual)	Anju Rawlley, Devdatta Vinayakrao Saraf
6	9789391505431	Applied Physics-I (with Lab Manual)	Vinod Kumar Yadav, Mina Talati
7	9789391505578	Applied Physics-II (with Lab Manual)	Hussain Jeevakhan
8	9789355387530	Basic Mechanical Engineering	Dharmendra Singh, Chandra Shekhar Rajoria
9	9789355387394	Building Construction	Rinku Kumar, Sandeep Panchal
10	9789355383280	Building Electrification	Arnab Ghosh
11	9789355384379	Building Planning and Drawing	Aparna Dwivedi
12	9789391505493	Communication Skills in English	Anjana Tiwari
13	9789355383570	Computer Aided Machine Drawing Practice	Kanak Kalita
14	9789355388964	Computer Networking and Data Communication	Sanjaya Shankar Tripathy
15	9789355382986	Computer Networks: Theory & Practicals	Brijendra Pratap Singh, Manoj Madhava Gore
16	9789355382672	Computer Programming: Theory and Practicals	Satyendra Singh Chouhan
17	9789355387714	Computer System Organization	Sonal Yadav
18	9789355382658	Concrete Technology: Theory and Practice	Sasankasekhar Mandal, Vishwajit Anand, Sushil K. Agarwala
19	9789355383334	Construction Materials and Testing	Vanita Aggarwal
20	9789355383501	Consumer Electronics	Amit M. Joshi, Maulin M. Joshi, Urvashi Prakash Shukla
21	9789355387684	Data Structures: Theory & Practicals	NB Venkateswarlu
22	9789355388513	Design of Machine Elements	A. Kumaravel, M. Kathirselvam
23	9789355388360	Design of Steel and RCC Structure	Sparsh Johari
24	9789355383303	Digital Communication	Muralidhar Kulkarni, K. S. Shivaprakasha
25	9789355387295	Digital Electronics	Menka Yadav
26	9789355389435	Electric Circuits and Networks	Jamuna K., Nilanjan Tewari
27	9789355388452	Electric Motors and Transformers (Theory and Practicals)	Sanjay B. Bodkhe
28	9789355385901	Electric Power Transmission and Distribution	Sudhir Sharma
29	9789355389466	Electrical & Electronics Measurements and Instrumentation	Sudarsan Sahoo, Vipin Chandra Pal, Sudipta Chakraborty
30	9789355382696	Electronic Communication	M. D. Selvaraj, P. Nagaradjane
31	9789355387202	Electronic Devices and Circuits	Ankesh Jain
32	9789355386373	Electronic Equipment Maintenance	Chanchal Sharma
33	9789355384690	Embedded Systems	Deepa M, G. Santhanamari, Susithra. N
34	9789355384867	Energy Conservation and Audit	M. Rizwan, Majid Jamil
35	9789391505509	Engineering Graphics	Sharad K. Pradhan, K.K. Jain
36	9789391505608	Engineering Mechanics (with Lab Manual)	Bhankhar Bharat, Gokaldas Vandana Somkuwar

DIPLOMA COURSE BOOKS

S.No.	ISBN	Book Name	Author
37	9789391505516	Engineering Workshop Practice	A. K. Sarathe
38	9789391505653	Environmental Science	Subrat Roy
39	9789355384966	Estimating, Costing and valuation	Sandeep Panchal
40	9789355389510	Fluid Mechanics & Hydraulic Machinery	Suman Chakraborty, Sourav Mitra, Aditya Bandopadhyay
41	9789391505592	Fundamentals of Electrical and Electronics Engineering (with Lab Manual)	Susan S. Mathew, Saji T. Chacko
42	9789355387240	Fundamentals of Power Electronics	Soumitra Kumar Mandal
43	9789355389602	Geotechnical Engineering (Theory & Practicals)	Neelima Satyam
44	9789355386137	Induction, Synchronous, and Special Electrical Machines	K. Selvajyothi, S. Tamilselvi, Sarat Kumar Sahoo
45	9789355382290	Industrial Engineering & Management	A. Manimaran
46	9789355384959	Internet of Things	Sujata Pal
47	9789355388650	Introduction to DBMS: Theory & Practicals	Myneni Madhu Bala
48	9789355384294	Introduction to e-Governance	Kamal Kant Verma
49	9789355386564	Introduction to Electric Generation Systems	Lalit Chandra Saikia, Nalin Behari, Deb Choudhury
50	9789391505585	Introduction to IT Systems (with Lab Manual)	Prashant Joshi
51	9789355382900	Linear Integrated Circuit	B. T. Krishna
52	9789355388148	Manufacturing Engineering	Santosh Kumar
53	9789355383877	Material Science & Engineering	Vinod Kumar, Nisheeth k. Prasad
54	9789391505424	Mathematics-I	Deepak Singh
55	9789391505523	Mathematics-II	Garima Singh
56	9789355385994	Measurements and Metrology	A. Gnanavelbabu, M. Santhana kumar, R. Saranraj
57	9789355388131	Mechanics of Materials and their Testing	R. Velmurugan, Pragathi
58	9789355385048	Microcontroller Applications	Ritula Thakur
59	9789355385062	Microcontrollers and Applications	Santanu Chattopadhyay
60	9789355388100	Mobile and Wireless Communication	Maulin Joshi, Urvashi P. Shukla
61	9789355387790	Production & Operations Management	Mayank Kumar Agrawal
62	9789355388704	Public Health Engineering	Rinku Kumar
63	9789355387455	Python Programming	Rupesh Nasre
64	9789355387189	Renewable Energy Technologies	Sanjay Agrawal, Rajeev K. Mishra
65	9789355387479	Software Engineering	Salman Abdul Moiz
66	9789355389749	Theory of Machines & Mechanisms	G. C. Mohan Kumar
67	9789355387332	Theory of Structures	Arunachalam Subramanian Balu
68	9789355389015	Thermal Engineering - I	V. Raghavan
69	9789355389831	Thermal Engineering - II	Parvesh Antil
70	9789355386779	Transportation Engineering (Theory & Practice)	A.K. Raji, K.K. Babu
71	9789355384287	Water Resource Engineering (Theory & Practice)	Prakash C. Swain, Dillip K. Ghose
72	9789355383051	Web Technologies: Theory & Practical	Raman Dugyala

CIVIL ENGINEERING SOCIETAL & GLOBAL IMPACT

AUTHOR

Dr. Shakuntala Acharya
Assistant Professor,
IIT Guwahati, Assam

REVIEWER

Dr. Nikhil Bugalia
Assistant Professor,
IIT Madras

KHANNA BOOK PUBLISHING CO. (P) Ltd.
4C/4344, Ansari Road, Darya Ganj,
New Delhi, India

All India Council for Technical Education
Nelson Mandela Marg, Vasant Kunj,
New Delhi, India

Dear Readers,

To prevent the piracy, this book is secured with HIGH SECURITY HOLOGRAM on the front title cover. In case you don't find the hologram on the front cover title, please write us to at contact@khannabooks.com or whatsapp us at +91-99109 09320 and avail special gift voucher for yourself.

Specimen of Hologram on front Cover title:

Moreover, there is a SPECIAL DISCOUNT COUPON for you with EVERY HOLOGRAM.

How to avail this SPECIAL DISCOUNT:

Step 1: Scratch the hologram
Step 2: Under the scratch area, your "coupon code" is available
Step 3: Logon to www.khannabooks.com
Step 4: Use your "coupon code" in the shopping cart and get your copy at a special discount
Step 5: Enjoy your reading!

Thanks,
Khanna Publishing House

ISBN: 978-93-5538-484-3

Book Code: UG429EN

CIVIL ENGINEERING SOCIETAL & GLOBAL IMPACT
by **Dr. Shakuntala Acharya**

First Edition: 2025

To view complete list of books, please scan the QR Code:

Published by:
Khanna Book Publishing Co. (P) Ltd.
CIN: U22110DL1998PTC095547

Visit us at: www.khannabooks.com
Write us at: contact@khannabooks.com

Printed in Bharat

प्रो. टी. जी. सीताराम
अध्यक्ष
Prof. T. G. Sitharam
Chairman

सत्यमेव जयते

आज़ादी का अमृत महोत्सव

अखिल भारतीय तकनीकी शिक्षा परिषद्
(भारत सरकार का एक सांविधिक निकाय)
(शिक्षा मंत्रालय, भारत सरकार)
नेल्सन मंडेला मार्ग, वसंत कुंज, नई दिल्ली-110070
दूरभाष : 011-26131498
ई-मेल : chairman@aicte-india.org

ALL INDIA COUNCIL FOR TECHNICAL EDUCATION
(A STATUTORY BODY OF THE GOVT OF INDIA)
(Ministry of Education, Govt. of India)
Nelson Mandela Marg, Vasant Kunj, New Delhi-110070
Phone : 011-26131498
E-mail : chairman@aicte-india.org

FOREWORD

Engineers are the backbone of any modern society. They are the ones responsible for the marvels as well as the improved quality of life across the world. Engineers have driven humanity towards greater heights in a more evolved and unprecedented manner.

The All India Council for Technical Education (AICTE), have spared no efforts towards the strengthening of the technical education in the country. AICTE is always committed towards promoting quality Technical Education to make India a modern developed nation emphasizing on the overall welfare of mankind.

An array of initiatives has been taken by AICTE in last decade which have been accelerated now by the National Education Policy (NEP) 2020. The implementation of NEP under the visionary leadership of Hon'ble Prime Minister of India envisages the provision for education in regional languages to all, thereby ensuring that every graduate becomes competent enough and is in a position to contribute towards the national growth and development through innovation & entrepreneurship.

One of the spheres where AICTE had been relentlessly working since past couple of years is providing high quality original technical contents at Under Graduate & Diploma level prepared and translated by eminent educators in various Indian languages to its aspirants. For students pursuing 2nd year of their Engineering education, AICTE has identified 88 books, which shall be translated into 12 Indian languages - Hindi, Tamil, Gujarati, Odia, Bengali, Kannada, Urdu, Punjabi, Telugu, Marathi, Assamese & Malayalam. In addition to the English medium, books in different Indian Languages are going to support the students to understand the concepts in their respective mother tongue.

On behalf of AICTE, I express sincere gratitude to all distinguished authors, reviewers and translators from the renowned institutions of high repute for their admirable contribution in a record span of time.

AICTE is confident that these outcomes based original contents shall help aspirants to master the subject with comprehension and greater ease.

T.G. Sitharam

(Prof. T. G. Sitharam)

Preface

This erring race of human beings dreams always of perfecting their environment by the machinery of government and society; but it is only by the perfection of the soul within that the outer environment can be perfected" (CWM, Aphorism 344)

Humans are abundantly creative! Modern wonders and ancient marvels exemplify this creativity and are testaments to the grit and tenacity of the great minds who have shaped our world today. However, unrelentless development sought at any cost is an erroneous dream and this book, titled 'Civil Engineering – Societal and Global Impact', is a reflection on and a response to, the profound need for sustainable development and the crucial role the designer- engineer plays in improving the quality of life and environment in which we live. The motivation of writing this book is to expose the civil engineering students to the varied dimensions of sustainability, the impact of civil engineering and its ability to be a vehicle of change for a sustainable future.

Keeping in mind the purpose of wide coverage as well as to provide essential supplementary information, the topics recommended by AICTE have been included in a very systematic and orderly manner throughout the book. Sections like multiple choice questions, short and long question and answers, and supplementary reading material with easy to refer QR codes have been provided, in text and in the 'Know More' Section. Various books, research publications, white papers and reports from national and international bodies have been incorporated. Apart from illustrations and examples as required, the book has been enriched with images from real locations, with in text explanations for proper understanding of the related topics.

In this first edition of 'Civil Engineering – Societal and Global Impact', an attempt has been made to cover the vast landscape of knowledge in the domain of Civil engineering, from its historical relevance and evolution, across technical know-how on infrastructure, the environment the built-environment, to pragmatic knowledge on codes, standards, and project management, through the lens of Sustainable development goals. All comments and suggestions that will contribute to the improvement of the future editions of the book, are welcome, and my sincere gratitude to all who enjoy this journey of learning. It has been a great learning experience for me as well.

The intent is to empower the students to be more than just an engineer, but a leader, a thinker, a problem solver; and imbibe in them a sense of social responsibility and environmental stewardship. I sincerely hope that this book will instil a sense of curiosity and awe in the world of Architectural design, Engineering and Construction, with Sustainability as a guiding principle, and encourage them to seek out creative and conscientious ways to contribute to society.

Dr. Shakuntala Acharya

Acknowledgement

I am grateful to the authorities of AICTE, particularly Prof. T. G. Sitharam, Chairman; Dr. Abhay Jere, Vice-Chairman; Prof. Rajiv Kumar, Member-Secretary, Dr. Ramesh Unnikrishnan, Advisor-II and Dr. Sunil Luthra, Director, Training and Learning Bureau for their planning to publish the books on 'Civil Engineering – Societal and Global Impact'.

I sincerely acknowledge the valuable contributions of the reviewer of the book,

Dr. Nikhil Bugalia, from Department of Civil Engineering, Indian Institute of Technology Madras for his constant support to bring this effort to fruition.

This book is an outcome of various suggestions of AICTE members, experts and authors who shared their opinion and thought to further develop the engineering education in our country. Acknowledgements are due to the contributors and different workers in this field whose published books, review articles, papers, photographs, footnotes, references, and other valuable information enriched us at the time of writing the book.

Dr. Shakuntala Acharya

Outcome Based Education

For the implementation of an outcome-based education the first requirement is to develop an outcome-based curriculum and incorporate an outcome-based assessment in the education system. By going through outcome-based assessments evaluators will be able to evaluate whether the students have achieved the outlined standard, specific and measurable outcomes. With the proper incorporation of outcome-based education there will be a definite commitment to achieve a minimum standard for all learners without giving up at any level. At the end of the programme running with the aid of outcome-based education, a student will be able to arrive at the following outcomes:

PO1. **Engineering knowledge**: Apply the knowledge of mathematics, science, engineering fundamentals, and an engineering specialization to the solution of complex engineering problems.

PO2. **Problem analysis**: Identify, formulate, review research literature, and analyse complex engineering problems reaching substantiated conclusions using first principles of mathematics, natural sciences, and engineering sciences.

PO3. **Design / development of solutions**: Design solutions for complex engineering problems and design system components or processes that meet the specified needs with appropriate consideration for the public health and safety, and the cultural, societal, and environmental considerations.

PO4. **Conduct investigations of complex problems**: Use research-based knowledge and research methods including design of experiments, analysis and interpretation of data, and synthesis of the information to provide valid conclusions.

PO5. **Modern tool usage**: Create, select, and apply appropriate techniques, resources, and modern engineering and IT tools including prediction and modeling to complex engineering activities with an understanding of the limitations.

PO6. **The engineer and society**: Apply reasoning informed by the contextual knowledge to assess societal, health, safety, legal and cultural issues, and the consequent responsibilities relevant to the professional engineering practice.

PO7. **Environment and sustainability**: Understand the impact of the professional engineering solutions in societal and environmental contexts, and demonstrate the knowledge of, and need for sustainable development.

PO8. **Ethics**: Apply ethical principles and commit to professional ethics and responsibilities and norms of the engineering practice.

PO9. **Individual and teamwork**: Function effectively as an individual, and as a member or leader in diverse teams, and in multidisciplinary settings.

PO10. **Communication**: Communicate effectively on complex engineering activities with the engineering community and with society at large, such as, being able to comprehend and write effective reports and design documentation, make effective presentations, and give and receive clear instructions.

PO11. Project management and finance: Demonstrate knowledge and understanding of the engineering and management principles and apply these to one's own work, as a member and leader in a team, to manage projects and in multidisciplinary environments.

PO12. Life-long learning: Recognize the need for and have the preparation and ability to engage in independent and life-long learning in the broadest context of technological change.

Course Outcomes

After completion of the course the students will be able to:

CO-1: Impact of Civil Engineering and Resource use

CO-2: Infrastructure and its resource requirements (energy, etc)

CO-3: Sustainability of the Environment, including its Aesthetics

CO-4: Potential of Civil Engineering towards Economy

CO-5: Built-Environment and factors impacting the Quality of Life

CO-6: Precautions and measures towards Sustainability

CO-7: Applying professional and responsible judgement, and developing a leadership role.

Course Outcomes	Expected Mapping with Programme Outcomes (1- Weak Correlation; 2- Medium correlation; 3- Strong Correlation)											
	PO-1	PO-2	PO-3	PO-4	PO-5	PO-6	PO-7	PO-8	PO-9	PO-10	PO-11	PO-12
CO-1	-	-	1	2	2	-	3	1	-	-	-	2
CO-2	1	1	1	1	1	2	3	-	-	-	-	-
CO-3	2	1	1	1	1	3	3	1		3	1	1
CO-4	-	-	-	-	-	-	3	2	3	1	3	-
CO-5	1	1	1	1	1	2	3	1	-	-	-	1
CO-6	1	1	1	1	1	2	3	1	1	1	-	1
CO-7	1	2	3	2	2	3	3	1	2	1	3	1

Guidelines for Teachers

To implement Outcome Based Education (OBE) knowledge level and skill set of the students should be enhanced. Teachers should take a major responsibility for the proper implementation of OBE. Some of the responsibilities (not limited to) for the teachers in OBE system may be as follows:

- Within reasonable constraint, they should manoeuvre time to the best advantage of all students.
- They should assess the students only upon certain defined criterion without considering any other potential ineligibility to discriminate them.
- They should try to grow the learning abilities of the students to a certain level before they leave the institute.
- They should try to ensure that all the students are equipped with the quality knowledge as well as competence after they finish their education.
- They should always encourage the students to develop their ultimate performance capabilities.
- They should facilitate and encourage group work and team work to consolidate newer approach.
- **They should follow Blooms taxonomy in every part of the assessment.**

Bloom's Taxonomy

Level	Teacher should Check	Student should be able to	Possible Mode of Assessment
Create	Students' ability to create	Design or Create	Mini project
Evaluate	Students' ability to justify	Argue or Defend	Assignment
Analyse	Students' ability to distinguish	Differentiate or Distinguish	Project/Lab Methodology
Apply	Students' ability to use information	Operate or Demonstrate	Technical Presentation/ Demonstration
Understand	Students' ability to explain the ideas	Explain or Classify	Presentation/Seminar
Remember	Students' ability to recall (or remember)	Define or Recall	Quiz

Guidelines for Students

Students should take equal responsibility for implementing the OBE. Some of the responsibilities (not limited to) for the students in OBE system are as follows:

- Students should be well aware of each UO before the start of a unit in each and every course.
- Students should be well aware of each CO before the start of the course.
- Students should be well aware of each PO before the start of the programme.
- Students should think critically and reasonably with proper reflection and action.
- Learning of the students should be connected and integrated with practical and real life consequences.
- Students should be well aware of their competency at every level of OBE.

Symbols and Abbreviations

LIST OF ABBREVIATIONS

Abbreviations	Full form
ADAM	Automated Drafting and Machining
AEC	Architectural design, Engineering and Construction
AI	Artificial Intelligence
AMRUT	Atal Mission for Rejuvenation and Urban Transformation
AR/VR	Augmented and Virtual Reality
ARAI	Automotive Research Association of India
ARCs	Autonomous Robotic Construction systems
AV	Autonomous Vehicles
BAS	Building Automation Systems
BEE	Bureau of Energy Efficiency
BIG	Bjarke Ingels Group
BIM	Building Information Modelling
BIS	Bureau of Indian Standards
BMS	Building Management Systems
BOM	Bill of Materials
BOQ	Bill of Quantities
BAS	Building Automation Systems
BEE	Bureau of Energy Efficiency
BIG	Bjarke Ingels Group
BIM	Building Information Modelling
CCS	Carbon Capture and Sequestration
CEOS	Committee on Earth Observation Satellites
CII	Confederation of Indian Industry
CLT	Cross-Laminated Timber
CM	Construction Management
COP	Climate Change Conference

CPS	Cyber-physical Systems
CSR	Corporate Social Responsibility
CTCN	Climate Technology Centre and Network
CTDP	Comprehensive Telecom Development Plan
C-DoT	Centre for Development of Telematics
C-V2X	Cellular Vehicle-to-everything
CAD	Computer Aided Drawing
CAR	Civil Air Regulations
DGCA	Directorate General Civil Aviation
DOE	Department of Electronics
DSD	Division of Sustainable Development
DSRC	Dedicated Short-Range Communications
DUAC	Delhi Urban Art Commission
ECV	Essential Climate Variable
EETSD	Environmental Education & Training for Sustainable Development
EF	Ecological Footprint
EMS	Energy Management Systems
ENIAC	Electronic Numerical Integrator and Computer
EoL	End of Life
EPA	US Environmental Protection Agency
EPI	Environmental Performance Index
ERNET	Education and Research Network
ETF	Enhanced Transparency Framework
EUNIS	European Nature Information System
EVA	Economic Value Added
EIA	Environmental Impact Assessment
EMS	Energy Management Systems
ENIAC	Electronic Numerical Integrator And Computer
EoL	End of Life
EPA	US Environmental Protection Agency
EPI	Environmental Performance Index
FAO	Food and Agriculture Organization
FRA	Forest Rights Act

GBCI	Green Business Certification Inc.
GDP	Gross Domestic Product
GEC	General Engineering Consultant
GHG	Greenhouse Gas
GIS	Geographic Information System
GoI	Government of India
GPS	Global Positioning System
GQII	Global Quality Infrastructure Index
GRI	Global Reporting Initiative
GRIHA	Green Rating for Integrated Habitat Assessment
HVAC	Heating, Ventilation and Cooling
HDI	Human Development Index
HGL-PCCB	High level Group for Partnership, Coordination and Capacity-building
HRIDAY	National Heritage city Development and Augmentation Yojana
i4.0	Industry 4.0
IAEG-SDGs	Inter-Agency and Expert Group on SDG Indicators
IAQ	Indoor Air Quality
IBC	International Building Code
IBoK	Indian Infrastructure Body of Knowledge
ICAO	International Civil Aviation Organisation
ICC	International Code Council
ICCROM	International Centre for the Study of the Preservation and Restoration of Cultural Property
ICOMOS	International Council on Monuments and Sites
ICT	Information and Communication Technologies
IEC	International Electrotechnical Commission
IEQ	Indoor Environment Quality
ITU	International Telecommunication Union
IUCN	International Union for Conservation of Nature
IWRM	Integrated Water Resources Management
IXP	Internet exchange points
IGBC	Indian Green Building Council
IHDI	Inequality adjusted Human Development Index

IIPDF	India Infrastructure Project Development Fund Scheme
INTACH	Indian National Trust for Art and Cultural Heritage
IoT	Internet of Things
IPCC	Intergovernmental Panel on Climate Change
IPD	Integrated Project Delivery
IS Codes	Indian Standard codes
ISHRAE	Indian Society of Heating, Refrigerating and Air- Conditioning Engineers
ISO	International Organization for Standardization
ISP	Internet Service Providers
LCA	Life-Cycle Assessment
LCC	Life Cycle Costs
LCEA	Life Cycle Energy Analysis
LCI	Life Cycle Inventory
LCIA	Life Cycle Impact Assessment
LEED	Leadership in Energy and Environmental Design
LiDAR	Light Detection and Ranging
LT-LEDS	Long-term Low greenhouse gas Emission Development Strategies
MoEF	Ministry of Environment, Forest and Climate Change
MoHUA	Ministry of Housing and Urban Affairs
NBC	National Building Code
NCSDCT	National Centre for Software Development and Computing Techniques
NDC	Nationally Determined Contributions
NHAI	National Highways Authority of India
NHPC	National Hydroelectric Power Corporation
NIOSH	National Institute of Occupational Safety and Health (NIOSH)
NRDW	National Rural Drinking Water Programme
NTPC	National Thermal Power Corporation
NUSP	National Urban Sanitation Policy
ODP	Ozone Depletion Potential
OECD	Organisation for Economic Co-operation and Development
OSH	National Occupational Safety and Health
OSHA	Occupational Safety and Health Administration
ODA	Official Development Assistance

PMC	Project Management Consultant
PPP	Public-Private Partnership
PV	Photovoltaic (panels)
QCI	Quality Council of India
QII	Quality Infrastructure Investment
RDSO	Research Designs and Standards organization
RICS	Royal Institution of Chartered Surveyors
ROI	Return on Investment
SDG	Sustainable Development Goal
SEA	Strategic Environmental Assessment
SFM	Sustainable facility management
SJVN	Satluj Jal Vidyut Nigam
SMART	Stormwater Management and Road Tunnel
SOM	Skidmore, Owings, & Merrill
SPCB	State Pollution Control Board
SPI	Smart Power India
SRI	Smart-readiness indicator
TEC	Telecom Engineering Centre
TOR	Terms of Reference
TPGEL	Tata Power Green Energy Limited
TSDSI	Telecommunications Standards Development Society of India
UN	United Nations
UNCED	United Nations Conference on Environment and Development (also, *Rio Conference or the Earth Summit*)
UNEA	UN Environment Assembly
UNEP	United Nations Environment Programme
UNCTAD	UN Conference on Trade and development
UNDESA	United Nations Department of Economic and Social Affairs
UNDP	United Nations Development Programme
UNSDSN	United Nations Sustainable Development Solutions Network
USGBC	US Green Building Council
UNESCO	United Nations Educational, Scientific and Cultural Organization
UNFCCC	United Nations Framework Convention on Climate Change

UNSC	United Nations Statistical Commission
UPI	Unified Payments Interface
VOC	Volatile Organic Compound
VDC	Virtual Design and Construction
WCED	World Commission on Environment and Development
WMO	World Meteorological Organization
WASH	Water, Sanitation and Hygiene
WHO	World Health Organization
WRM	Water Resource Management

List of Figures

Unit 1: Impact of Civil Engineering: An Introduction

Unit 2: Importance of Civil Engineering

Unit 3: Infrastructure

Unit 4: Environment

Unit 5: Built Environment

Unit 6: Civil Engineering Projects

Contents

01 IMPACT OF CIVIL ENGINEERING: An Introduction

UNIT SPECIFICS

Through this unit we have discussed the following aspects:

- ***The past to look into the future:***
  - *Pre- industrial revolution*
  - *Agricultural revolution*
  - *first and second industrial revolutions*
  - *IT revolution*
- ***Major Civil Engineering breakthroughs and innovations–a Timeline***
- ***Sustainable Development–Present day world and future projections***
  - *The steady erosion in Sustainability and its global impact*
  - *Sustainable Development Goals (SDGs), its impact and possible causes*
- ***Evaluating future requirements for various resources***
  - *GIS and applications for monitoring systems*
  - *Human Development Index and Ecological Footprint*

Besides giving a large number of multiple-choice questions as well as questions of short and long answer types marked in two categories following lower and higher order of Bloom's taxonomy, a list of references and suggested readings are given in the unit so that one can go through them for practice.

There is a "Know More" section, which has been carefully designed so that the supplementary information provided in this part becomes beneficial for the users of the book. It is important to note that for getting more information on various topics of interest some QR codes have been provided which can be scanned for relevant supportive knowledge This section mainly highlights applications of the subject matter for our day-to-day real life or/and industrial applications on variety of aspects, case study related to environmental, sustainability, social and ethical issues whichever applicable, and finally inquisitiveness and curiosity topics of the unit.

RATIONALE

This introductory unit on the historic and socio-economic drivers behind the emergence of Civil engineering and its profound impact on sustainable development sets stage for the chapters ahead.

UNIT OUTCOMES

List of outcomes of this unit is as follows:

U1-O1: Understanding/ comprehension of world history and socio-economic context U1-O2: Knowledge on the major breakthroughs and innovations of the domain

U1-O3: Knowledge on Sustainable development, its Goals, Impact and Indicators

Unit-1 Outcomes	EXPECTED MAPPING WITH COURSE OUTCOMES (1- Weak Correlation; 2- Medium correlation; 3- Strong Correlation)						
	CO-1	CO-2	CO-3	CO-4	CO-5	CO-6	CO-7
U1-01	2	2	3	1	2	1	
U1-02	3	1	2		2	1	
U1-03	3	2	3		2	3	2

The distinct most character of a civilisation is the advent of organised social and physical infrastructures to improve the quality of civic life. Though formal recognition of this branch and degree first appeared in the early 18th century, the practice and profession of civil engineering has existed since the dawn of civilisation. It is the branch of engineering or professional discipline of; planning, designing, building/construction and maintenance of physical infrastructures and built environment, to serve the general public or civilians. Since the designed built environment is a testament to the cultural heritage of a civilization, and a reflection of the socio-cultural, scientific, and technological advancements of the time, Civil engineering plays a pivotal role in impacting the society and global population.

In the following Section, the historical events that transformed society and challenged the discipline of civil engineering to develop is discussed.

1.1 DAWN OF CIVIL ENGINEERING: A BRIEF HISTORY

1.1.1 Pre-Industrial Revolution

Early human history can be dated back to the Stone Age in the Palaeolithic period (2.5 million years ago – till 10,000 BCE) when early man followed nomadic ways of the hunter-gatherer, lived in caves and huts or *tepees* and had begun to develop rudimentary tools of stone and wood. But around 10,000 BCE, a shift in climate and discovery of farming led the Neolithic man to settle, mostly along riverbanks, with focus on production of food. This in turn, led to the need to build secure habitation, plan settlements, and land use, construct transport and irrigation systems, and design solutions towards water supply and sanitation, which can be identified as the first examples of civil engineering.

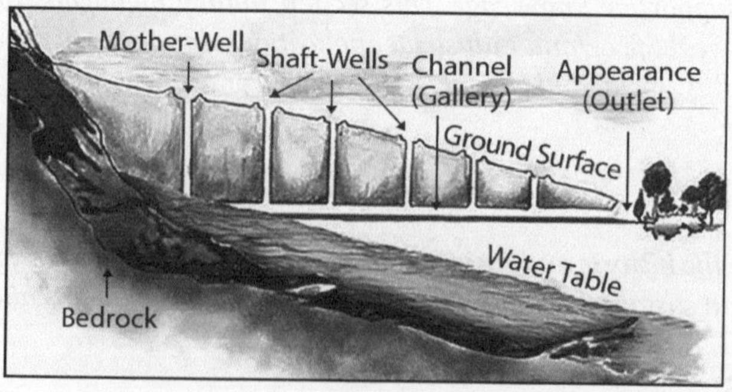

Fig.1.1: Persian Qanat Profile (source: Manian, et al., 2022)

In the following era, known as Bronze age, the social organisation altered with kings and leaders emerging and early towns being established, leading to the migration of people for better wages and employments. Skills, tools, and techniques that were handed down from generation to generation of skilled artisans, stonemasons, and carpenters, in a small locale developed into expertise and artistry that led to the rise of planners, architects and engineers, tackling projects traversing regions and empires. Between 4000 and 2000 BC, the profession grew from being utilitarian to being a creative practice of achieving feats of grandeur as is exemplified in the intricate planning of urban housing, sanitation, and water systems in the Indus Valley civilisations; decadent pyramids and monuments of Egypt; *Qanats* in Persia and Mesopotamia; Stonehenge in UK and the Great Wall of China; where functionality and aesthetics paralleled each other. *This is discussed further, with detail in the next Unit.*

Most of the construction efforts were achieved by manual labour and non-mechanised tools, however, two major global occurrences, namely, the Agricultural Revolution and the Industrial Revolution, created a huge demand of infrastructure for, housing, transportation, sanitation, and water management, and eventually for environmental planning and construction management.

1.1.2 Agricultural Revolutions

The **First Agriculture Revolution**, also known as the Neolithic revolution, occurred in 10,000 BC and led to the organisation of modern man as a producer, rather than a hunter- gatherer. While the following millennia focussed on socio-economic development and in turn, creative feats of engineering and design, as discussed above; agriculture continued to be the primary occupation of the common folk. Across all the major civilisations that arose, from those in Indus Valley, Egypt, Mesopotamia, Greece, Rome, Mesoamerica, Europe, Arabia and eventually the new nations and colonies; innovations and developments centred on the methods, tools and techniques to support agriculture, through construction of water supply for irrigation and sanitation, granaries for food storage, tools and processes for production and preservation, etc.

However, by the mid- 17th century, an unprecedented growth in agricultural production from increase in labour and land productivity was observed in two nations, Britain and Netherlands. This was notably the **Second Agriculture Revolution**, and it spread across the other nations in Europe and their colonies in East Asia and North America over the next two centuries until the late 19th century. While the Dutch observed increase in the agricultural output per labourer, the British experienced the abundant outgrowth of output with respect to the population.

The strategic developments and innovations of the British Agriculture Revolution that shaped the era were; introduction of new crops, selective breeding and the method of crop rotation; improved design of the plough called the Dutch swing plough; and institution of land reforms, such as, land conversion, drainage and reclamation, increase in farm size, and enactment of Enclosure Act. Civil engineering played a significant role here, in the development of transportation infrastructure – roads, canals, railways, as well as social infrastructure, such as, development of national markets, supplemented with policies free of tariff, tolls and customs.

It is interesting to note that, while abundant food supply led to rampant increase in population, it eventually led to loss of labour force engaged in agriculture, leading to a need for new occupations. With time, though the inland production was countered with cheap imports of food supply, this revolutionary spike ensured a complacent food reserve that allowed a shift in the National priorities, from agriculture to industry. A new working class, seeking urban employment in the upcoming industries, emerged as a result and fuelled the onset of the Industrial Revolution.

The first half of the 20th century was rife with war and caused major setbacks in the developed countries. It is also at this time that several colonised nations sought their independence, further adding to the need for re-building and self-sufficiency.

The **Third Agriculture Revolution,** popularly known as the **'Green Revolution',** transpired as a response to global hunger and poverty, dilapidated infrastructure, and scarce and contaminated natural resources by mid-20th century. New technologies and cutting-edge research in the areas of high-yield crops, hybridized seeds, genetically modified organisms (GMOs); use of chemical fertilizers and pesticides; methods of controlled irrigation and mechanized cultivation; characterise this era.

MS Swaminathan, agronomist and parliamentarian, loving called the '*Wheat man of India*' invited Norman Borlaug, the 'Father of the Green Revolution', at the brink of a famine in 1961, and initiated collaboration with the Ford Foundation to import newly developed variants of wheat and later rice. Punjab became the birthplace of the **Indian Green Revolution or 'Harit Kranti'** and the Government of India facilitated programs to support farmers in the use of agrochemicals and irrigation. Late Prime Minister Shri Lal Bahadur Shastri, gave the slogan *'Jai Jawan Jai Kisan'*. Today, India is the world's largest rice exporter.

Fig.1.2: '*Harit Kranti*'–The Indian Green Revolution

Presently, several strategies and contributions of the 'Green revolution' is being questioned, despite the fact that it addressed world hunger, poverty, and reconversion of land for agriculture. The decrease in food security and increase in production of export crops, has been unwarranted results of the same. However, while a rehauling of perspectives is underway to ensure long-term sustainability, some of the fundamental innovations and engineering contributions of this time set the tone for the future.

1.1.3 Industrial Revolution

While the North American colony was the primary producer of cotton, India was compelled to cultivate indigo on large-scale, despite several revolts by farmers. These two crops propelled the production of indigo-dyed cotton fabric, which in turn, thrust the need to improve the existing hand production to machines and mills, leading to the evolution of the design of cotton mill–from *Spinning Jenny* (1764) to the *Power Loom* (1787), to meet growing demands and establish a monopoly in international trade. This marks the **First Industrial Revolution**, spanning approximately 100 years from 1760 – 1850, characterised by mechanisation, fuelled by abundant coal supply and the optimisation of the design of steam engine by James Watt in 1765. new profession in engineering – mechanical engineering, was born, and several machine and production line innovations were developed.

Studies and experiments looking into the phenomena of electricity, by Benjamin Franklin, James Watt and Alessandro Volta, eventually gave rise to the **Second Industrial Revolution** between 1850 – 1917 with electrical energy as a new source of power that enabled mass production. However, by the mid-19th century, the working conditions in the factories across Europe worsened and there was worker unrest leading to The Great Reform Movements, thereby slowing the industries. By early 20th century, World wars and demand of independence by colonies, further led to the slowing of Europe as a forerunner in industry, and USA emerged as the promised contender.

The span between 1877-1917, termed as the **'Gilded Age'** of America by satirical author, Mark Twain, is characterised by massive economic growth, fuelled by development in the areas of manufacturing, railroads, telecommunication, automotive, product and industrial design, in the newly united federate, post-civil war. A centralised government and growing bureaucracy looked at expanding their jurisdiction and focussed on public services, through building of extensive rails and roads, telecommunication infrastructure, **'mass production'** industries and support facilities for public health and sanitation. A new connected world gave a global overview to markets, businesses, politics, and the invention of the lightbulb by Edison allowed prolonged working hours, furthering the industrialisation agenda.

The role of engineers, sanitarians and public health scientists grew in importance, while new domains of **product design and industrial design** came into being. Some of the common products that are used till date, came into conception during this era, such as, the foot- powered sewing machine, patented by Isaac Singer in 1851; the electric iron, invented by Henry Seeley in 1882; the first drum washing machine, patented by James King in 1851; and the first mechanical, hand-cranked, dishwashing device made of wood, was registered in 1850 by Joel Houghton.

Another key area of innovation continued to be **telecommunication**. Graham Bell invented the telephone in1876 and founded the AT&T company. While Edison invented the phonograph–a sound recorder, in 1877 and the kinetoscope–a peephole video player, in 1892.

Transportation remained a focus and a remarkable innovation that took place at that time was the design of the 'Quadricycle' by Henry Ford in 1896, that featured four bicycle wheels powered by a four-horsepower, internal combustion engine fuelled by petroleum and not steam. This led to the establishment of the Ford Motor Co. in 1903, which became the America's largest car manufacturer, housing a moving assembly line (*refer Fig. 3*) that made the Ford model T, a household name. That same year, the Wright brothers took first flight in the powered airplane, the 1903 Wright Flyer.

Fig.1.3: Assembly Line at Ford Motor Co. *(source : www.corporate.ford.com)*

The use of ***new source of power – electricity and petroleum****; new materials– iron, steel, glass, rubber; new techniques and design of engineered systems – turbines, engines, motors, conveyors, etc.; new technology, such as, Babbage's Analytical Engine or the first computer, led to mass-production, eased transportation and improved computation and telecommunication. Unfortunately, the early 20th century saw the onset of two World Wars, during which, while great scientific and engineering progress was made, catastrophic societal and environmental damage was onslaught.*

1.1.4 Digital Revolution and Industry 4.0

An era of rebuilding began around 1947, which is noted as the Third Industrial Revolution or Digital Revolution, characterised by the adoption of digital technologies, many of which had its precursors developed during the World Wars. It is also at this crucial time in history when India received her independence and was faced with many challenges, and took time to catch up to the state of the art.

The early years of the '*Information Age*' saw several firsts. In 1943, the ENIAC (Electronic Numerical Integrator And Computer) was developed; In 1947, the first working transistor was designed at Bell Labs, and later the MOFSET or MOS transistor was developed. In parallel, at Fairchild Semiconductor; the first monolithic integrated circuit chip was developed in 1959, alongside research and development in the area to improve the MOS chips, paving the way for the first microprocessor, Intel 4004. This technology also led to the development of image sensors for future digital cameras. In the 1960s and 70s, early CAD (Computer aided Drawing) software were sprouting, with Sutherland's SKETCHPAD, popularly nicknamed, 'Robot Draftsman' and Hanratty's Automated Drafting And Machining (ADAM).

1969 saw the breakthrough invention of the 'internet', when a message over the first wide-area network, ARPANET, was sent to public. This further led to development of inter- networking protocols, and with the introduction of the first home computer in 1970, a proliferation of digital technology and information sharing became the new way of life. Apart from utilitarian needs of digital record keeping, computation and automation, computers also led to a thriving video gaming industry that excelled in the development of first game consoles, game graphics and arcade games. Over the next decade, industrial robots, CGI in film and animation, electronic music and signages, were widely incorporated and eventually, the first mobile phone by Motorala in 1983 and the first digital camera in 1988 came into being.

The internet arrived in India in 1989, through the development of our indigenous network commissioned by the Department of Electronics (DoE), modelled on the ARPANET, named ERNET (Education and Research Network) and as a start, it connected IISc Bangalore, 5 IITs, NCSDCT (National Centre for Software Development and Computing Techniques) and DoE. Hardware and computers were already being imported by 1981 and India was already exporting software. In 1984, C-DoT (Centre for Development of Telematics) was established under the leadership of Sam Pitroda to design digital exchanges, but later expanded to develop software applications. In 1993, satellite link became operational in Bangalore Software Technology Park, and by 1998, VSNL launched Internet services that enabled companies operating from these parks to engage international clients. The Information Technology India market accounts for 9.3% of India's GDP and 56% of the global outsourcing market (IBEF, 2022) and all six Indian brands feature among the top 10 fastest-growing IT Services brands over the course of 2020-2022.

While digitisation remains integral to our present day lives, a new technological advancement is on the horizon, termed as **Industry 4.0 or the 4th Industrial Revolution,** characterised by artificial intelligence, advanced robotics, internet of things (IoT), cloud computing and cyber-physical systems (CPS). 'i4.0', coined in 2015, leverages interconnectivity and big data to facilitate quick and decentralised decision-making and allows to cater to customised solutions for individual needs, in contrast to 'mass -produced' solutions, and imbibes increased automation, improved communication, Machine-to-machine as well as Human-Computer Interactions, and self-monitoring and autonomy.

Several i4.0 technologies, such as, Building Information Modelling (BIM) with cloud technologies, additive manufacturing or 3D printing, AR/VR (Augmented and Virtual Reality), drones and unmanned aerial vehicles, etc. is widely being used in the Architectural design, Engineering and Construction (AEC) domain today.

1.2 MAJOR BREAKTHROUGHS AND INNOVATIONS

From early man's tools to present technology of using drones, unmanned vehicles, and cloud computing; the major breakthroughs and innovations that moulded what civil engineering is today is as follows:

Tools, the first breakthrough that enabled construction, were developed from available natural materials, dating back to the Neolithic age. Stone axe with wooden handle, bone hammer, stone adze, celt, flake tools, sickle, drills and even, sledge, have been discovered from that era. Ancient tenements and structures were primarily built of raw natural resources that were abundant in the region, such as, rocks and stones, timber and bamboo, mud, and clay, etc.

Bricks, were a major innovation and till date, is the integral building-block of most structures. Sun-baked mud bricks or '*adobe*', cemented by lime mortar, was extensively used in Indus Valley and Egypt. Later, kiln-fired, and glazed bricks were developed in Mesopotamia. Large rocks were also chiselled and faced into massive bricks. This in turn, led to innovations in new tools, such as, ramp, lever, lathe; and techniques, such as, butterfly interlocking, method of drilling stone, enamelling; led to development of complex and tall structures, such as, pitched- brick vaults, and pyramids.

Cranes, pulleys, and jibs to raise construction materials to great heights, employing metal cramps to join large stone blocks, and development of early 'construction drawings', to build mega-structures, such as, groin vaults, arch bridges and the early multi-storied buildings were some of the contributions of the Greek.

Road design and construction, as we know it today, was first formidably developed by Romans **and the first known roadway, the Appian Way** or the '*Queen of the roads*', was constructed in 312BCE, connecting Rome with its allies in Capua. The design comprised of digging shallow trenches with retaining walls on either side, filled with layers of; levelled earth and mortar or sand topped with rocks, crushed rocks or gravel cemented with mortar and surfaced with blocks of cut rocks, arranged pebbles, iron ore or hardened lava.

Roman cement, developed by adding volcanic ash, called *pozzolana*, in lime mortar to make it harden under water, is one of the major breakthroughs in the history of civil engineering and construction. This further led to the development of Roman concrete, where concrete made of rubble and mortar was filled inside stone or brick formwork, which was later replaced with the use of removable wooden shuttering; this technique allowed building of arches, barrel vaults and domes over large spans.

Romans also introduced, the use of lead or '*plumbum*' for pipes and roof covering, thus being the origin of the word plumbing; and centralised heating achieved through raising the floor above a wood or coal fire exhaust. They also made use of glass for windows and in mosaics.

Iron reinforcement dates back to the time of the ancient times when cramps and bolts were in use to either hold together stone blocks or timber members in trusses. Later in the 15th century, the same were used by Brunelleschi in the design of the dome, where two layers of domes met at the top in an open stone compression ring and iron cramps held together tie rings running horizontally between ribs. The design was improved in cupola of St. Peter's Basilica, where three continuous iron chains held the dome in tension.

Pile driver, a tool to drive 'piles' or vertical, pole-like, structural members of deep foundation, driven deep into the ground where the soil is loose, to support piers, bridges, cofferdams, etc., was invented in 1500.

First railway line has been referred to in 1515, having wooden rails and a hemp haulage rope through a treadwheel operated by manual power. Several funiculars have been spotted across Europe, however, the world's oldest operational railway–the Middleton Railway, was built in 1758 in Leeds, UK.

Drafting and surveying tools like, the line gauge, plumb line, the carpenter's square, the spirit level, and the drafting compass were developed in the 17th century.

Structural iron was introduced as a building material with the availability of iron, soon replacing wood and charcoal. Iron ore was smelt with the use of coke on a mass scale in 1702. Christopher Wren used iron hangers to suspend floor beams at Hampton Court Palace, and iron rods to repair Salisbury Cathedral and strengthen the dome of St Paul's Cathedral, and iron columns in the House of Commons. Cast-iron was used for the bridge at Coalbrookedale in 1776, while wrought iron was used for the roof structure of the Louvre, Paris. Later in the 19th century, the first two exhibitions of The **Great Exhibition of the Works of Industry of All Nations**, or the *Exposition Universelle*, namely, the Crystal Palace in Hyde Park, London, and the Eiffel Tower, Paris, were also exemplars of iron structures.

Modern road design developed in the 18th century, with method improvements and new technique proposals by Tresaguet, Metcalf, Telford, and McAdam. The present design is an adoption, named the '*Tar McAdam Road*', patented in 1901.

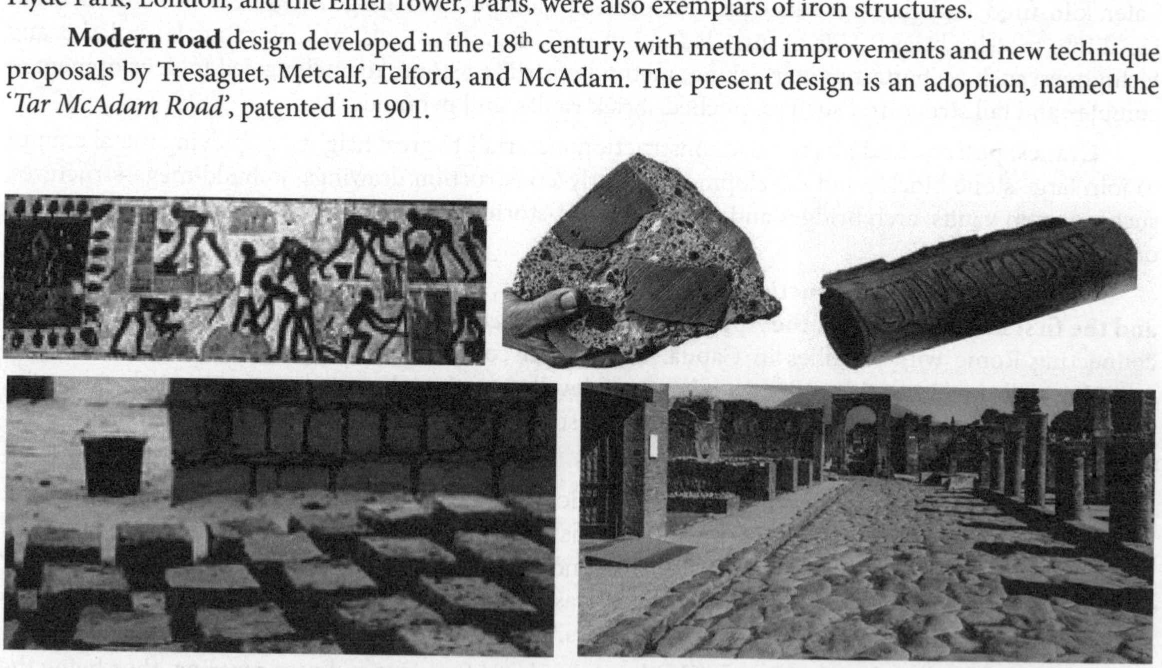

Fig.1.4: Ancient building materials and techniques–(*Top Left*) Egyptian Hieroglyph showing building construction with brick; *(Top Middle)* Roman Cement, (*Top Right*) Lead 'plumbum' Pipes, (*Bottom Left*) Sun-dried Mud bricks of Mesopotamia ; (*Bottom Right*) Roman Road

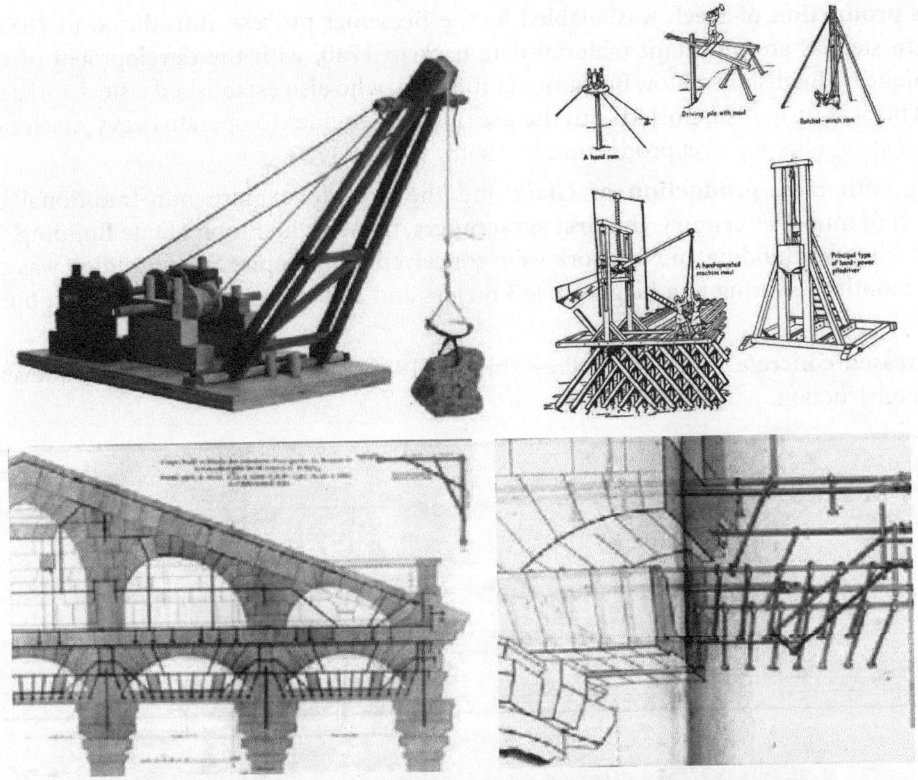

Fig.1.5: Ancient construction tools and techniques–(*Top Left*) Greek crane, (*Top Righ*) Designs of Pile Driver, (*Bottom*) Rondelet's drawing of reinforcement iron bars (Blasi and Iori, 2008)

First iron chain suspension bridge in the Western world was the Jacob's Creek Bridge (1801), Westmoreland County, Pennsylvania, designed by inventor James. However, Union Bridge (1820) is the oldest operational chain suspension bridge.

The **first wire-cable suspension bridge** was the Spider Bridge at Falls of Schuylkill (1816),

Portland cement was patented in 1824 by Joseph Aspdin and its modern common version was developed by his son, William, in the 1840's.

The **first** advertised **prefabricated home** was the "Manning Portable Cottage" conceived in 1830 by London carpenter H. John Manning. Shortly after in 1845, Isambard Brunel designed the prefabricated wood and canvas, Renkioi Hospital, for assemblage on site at the Crimean War.

The **first systematic national building standard** was established with the London Building Act of 1844, and by 1855, Metropolitan Board of Works was set up. Meanwhile in the USA, the City of Baltimore passed its first building code in 1891and by 1908 a formal building code was rafted and adopted.

Reinforced concrete was invented in 1849 by Joseph Monier and the first reinforced concrete bridge was built by the inventor-engineer.

Building of the **Panama Canal** between **1905-14, was** a feat of civil engineering as it was the biggest earth dam in the world and created the largest man-made lake, by deepening of the Pacific and Atlantic **canal** entrances, widening and deepening the Gatun Lake navigational channel. It remains one of the busiest shipping lanes in the world.

Mass production of Steel, was enabled by the Bessemer process, introduced in 1855. However, the move to steel as an important material date back to 1740, with the development of the crucible steel technique by English inventor, Benjamin Huntsman, who also established a steelworks at Sheffield, England. This was further supported with the use of steam engines to operate heavy machinery and the invention of steel roller for steel production by Henry Cort in 1783.

Along with **mass-production of Glass** and the urge to explore non-traditional design and construction in mid- 19th century, **the first skyscrapers**, namely, the Empire State Building, Rockefeller Center and Chrysler Building, in New York were conceived. The Empire State Building was constructed in only 13 months, towering to a height of 443 meters and 102 stories, it was the tallest building of its time in 1931.

Prestressed concrete invented by Freyssinet in 1928 and further applied it to develop precast segmental construction.

Fig.1.6: Panama Canal (*Top*) A Cross Section Canal, (*Bottom*) Picture

Fig.1.7: Modular construction–Habitat 67, Montreal *(Left)* Section drawing, (*Right*) Photograph

The following innovations are of particular importance in modern construction and will be further discussed in later Units.

Modular construction of home systems first materialised in 1933 with the Winslow Ames House by Robert W. McLaughlin and his firm, American House Inc. Cemesto, a panel board made from sugarcane, patented by the John B. Pierce Foundation, was used and in 1942 the US government employed Skidmore, Owings, & Merrill (SOM), to come up with a scheme called "Flexible Space". One of the most remarkable prefabricated, modular megastructures remains the Habitat 67, by Moshe Shafdie for the Expo in Montreal. The market is projected to increase from $92.18 billion in 2018 to $130 billion in 2030.

Development of CAD (Computer Aided Drawing) software can be traced back to the development of PRONTO, the first commercial numerical-control programming system by Dr. Patrick Hanratty in 1957, and to the first 'Robot Draftsman' or SKETCHPAD application developed by Ivan Sutherland's during his doctoral work at MIT in 1963, that used a GUIA graphical user interface that facilitated human-computer interaction through visual aids or icons.

BIM (Building Information Modeling) as a concept can be traced between 1970-1990 to the first software tools developed for modelling buildings, i.e., the Building Description System, GLIDE, RUCAPS, Sonata, Reflex and Gable 4D Series. The term 'Building Information Model' first appeared in a 1992 paper authored by van Nederveen and Tolman, but it was only in 2003 that Jerry Laiserin, acknowledged the contributions of Autodesk, Graphisoft and Bentley systems and standardized the term as a common name for the digital representation of the building process.

Construction 3D Printing is the technology of using additive manufacturing technique through computer-controlled activities of; sequential extrusion of material, such as, 3D concrete, powder bonding and additive welding, through Autonomous Robotic Construction systems (ARCs) or Freeform Construction. While empirical development of this technology started out in early 2000s, it may be traced back to the 1950's when robotic bricklaying and on- site automated fabrication was being explored. Presently various applications, such as, fabrication of houses and construction components, building bridges and canals, and even artificial reefs are being made (*Refer KNOW MORE Section to know about India's Research in this area*)

Smart buildings and Building Automation systems are terms used in conjunction today to refer to buildings that use IoT technology to monitor and connect various devices, sensors, hardware, software to manage various building services, such as, HVAC, lighting, fire protection, security, and access control, etc.

While various innovations have transformed design, engineering and construction, good old-fashioned soundness of the principles of civil engineering remains fundamental to the success of a built- environment and the hallmark of a good engineer. This is elucidated by the infamy of the Tower of Pisa, popularly called the ***'Leaning Tower'***, near Florence Italy, and the remarkable team that saved it from collapsing. The 56m freestanding *campanile*, or bell tower, began to lean within five years of work commencing in 1173, caused by shallow foundations in unstable sub-soil. It took 200 years to complete, all the while continuing to tilt, until 1993 when it was reportedly at an alarming 5degree tilt. This lean was corrected by a team of experts that devised a solution called 'soil extraction', where a series of tunnels on the side opposite to the tilt was dug to drain the soil and remove small amounts of soil, while reinforcing the foundations with 15m concrete pillars and harnessing with steel cables to pull it back in position, till it self-corrected by 2001.

Every breakthrough and innovation, however, has profound consequence associated with it. These impacts maybe positive, like development of civic amenities, transport facilities, telecommunication, and global trade; or negatives, like deforestation, loss of indigenous flora and fauna, colonisation, poor labour conditions and poverty; and are increasing at a global level with the advent of globalisation and enhanced connectivity. Civic life is nested on Society, Economy, and Ecology (or Environment), known as the three pillars of Sustainability, with the overarching intent to improve quality of life for today's populace and future generations.

1.3 SUSTAINABLE DEVELOPMENT: PRESENT DAY WORLD AND FUTURE PROJECTIONS

With Industry 5.0 on the horizon, a long-term, sustainable point of view for development is the need of the hour. The concept of 'Sustainability' was first introduced in the context of development in the book '*Our Common Future*', popularly known as the Brundtland Report, from the World Commission on Environment and Development (WCED), published in 1987 by the United Nations. It defined the term as "*meeting the needs of the present without compromising the ability of future generations to meet their own needs*" and propagated the principles of 'Sustainable development' that focuses on environmental protection and social equality along with economic growth. This eventually laid the foundation for formulating global Sustainable Development Goals (SDGs) in 2015, adopted by 193 committed member states of the United Nations with the projection of meeting the goals by 2030.

1.3.1 The steady erosion of Sustainability and Mitigative Actions

Following the world wars and the advent of the Third Agricultural revolution, two evocative pieces, namely, '*The Silent Springs*' (Carson, 1962) which argued against excessive use of pesticides like DDT and its harmful impact on several species, and '*Scarcity and Growth*' (Barnett and Morse, 1963) which empirically established the burden on natural resources, can be noted as the beginning to a paradigm shift at a global level to address issues of sustainability at the intersection of social and environmental or natural ecosystems. However, the impending ecological crisis was first noted in the previous century, with Swedish scientist Arrhenius, in 1896, predicting the change in surface temperature owing to greenhouse effect caused due to increased fossil fuel use, later corroborated by Guy Callendar in 1938 who connected carbon dioxide increase in earth's atmosphere to global warming. Research across 1940's through the 1960's revealed the implications of CO_2 emissions; with Plass reporting that CO_2 in atmosphere captures infrared radiation, otherwise lost to space, and Keeling producing concentration curves for atmospheric CO_2 showing a downward trend in annual temperatures. The 60's was a decade of ecological strife, with Monaco opposing French plan to dump radioactive waste in the Mediterranean Sea, and catastrophes such as, the Torrey Canyon oil tanker spillage off the coast of England and the Santa Barbara oil spill. These findings and events led to the first Earth Day celebration in 1970 and the creation of the US Environmental Protection Agency (EPA). Further, an inter-governmental conference of experts for Rational use and conservation of Biosphere (UNESCO, 1970) was convened, and in 1990 **Earth Day** went global.

India, too, saw its share of conservation movements. The *Chipko Movement* of 1973 was a non-violent social and ecological '*andolan*' by rural villagers, mostly women, protesting government backed logging in the Himalayan regions of Uttarakhand, India. Inspired by the Bishnois of Khejari, who were killed trying to protect their sacred trees in 1730's in the kingdom of Marwar, the Chipko protestors stood hugging trees and in 1987, this movement was recognised with the Right Livelihood Award for its conservational efforts. This further inspired several other similar conservation movements across Rajasthan, Bihar, Himachal Pradesh, the '*Appiko movement*' in Karnataka and the Western Ghats in 1983, as well as campaigns to protect the canopies planted along the Grand Trunk Jessore Road in West Bengal, as recent as 2017. India enacted the *Wildlife (protection) Act* in 1972, *Water (Prevention and Control of Pollution) Act* in 1974, *Air (Prevention and Control of Pollution) Act* in 1981, followed by the *Environment (Protection) Act* in 1986. The 21st century has seen several amendments and additions, such as, special regulations to Environment Act for ozone-depleting substances in 2000 and Coastal zone notification in 2018, The *Energy Conservation Act* 2001, and *Biological Diversity Act* 2002. While most of these aimed at protection and conservation of natural Environment, in 2006 the enactment of the Forest Rights Act (FRA) for scheduled tribes and other traditional forest dwellers, recognised the interdependency within the socio- ecological ecosystems and introduced an unprecedented reform at the intersection of the Ecology and Social dimensions of sustainable development.

While development remained the key motivation, various strategies, policies and acts were established between 1967- 1987 to further the sustainability discourse in the West. The *Clean Air or Air Quality Act* (1967), the *National Environmental Policy Act* (PEA, P.C.E.A., 1969) and the *Endangered Species Act* (1973) were passed by the US Congress; and legal action for environmental damages was pursued by Environment Defense Fund (EDF 1967, www.edf.org). Several books and papers, such as, '*Only One Earth*' (Dubros and Ward, 1971), '*Limits to Growth*' (Meadows et al., 1972), '*Polluter Pays Principle*' (OECD, 1972) , '*World Conservation Strategies*' (Talbot, 1980, released by IUCN), '*Global 2000 Report*' (Barney, 1980 and Council on Environmental Quality), and UN's '*World Charter for Nature*' (un.org) were published; and various events, such as, '***Habitat***'–the UN Conference on Human Settlements in 1967, the International Conference on Environment and Economics by OECD in 1984, and the UN General Assembly of 1987, where the foreword of '*Our Common Future*' was presented, took place. This culminated into the '*Agenda 21 Declaration of Environment and Development*' (UNCED, 1992), an action plan proposed for the 21st century in the 1992 Rio Earth Summit. The UN Conference on Sustainable Development or ***Rio+20*** was held twenty years later in 2012, where the process to develop measurable goals and targets as a set of Sustainable Development Goals (SDGs) was adopted, along with green economy policies, and in 2015 the fifteen-year plan to '*2030 Agenda for Sustainable Development*', bearing the 17 goals and 169 targets were committed to by 193 member nations. Currently, between October and December 2022, the Rio +30 conference is being organised.

1.3.2 Sustainable Development Goals and Global Impact

The present condition, post pandemic, with respect to each SDG (sdgs.un.org) is as follows:

1. **No Poverty** – COVID 19 pushed 8 million workers into poverty worldwide and working poverty rate climbed for the first time in two decades. Furthered by inflation, wars and political crisis, and disaster there is an ongoing migration and refugee crisis making ending poverty the foremost priority.
2. **Zero Hunger** – 1 in 10 people worldwide are suffering from hunger and 1 in 3 people lack regular access to adequate, in addition, food shortages and soaring food prices affected 47% of countries in 2020, up from 16% in 2019. Ending hunger, achieving food security and improved nutrition, and promoting sustainable agriculture are some of the primary targets.
3. **Good health and well-being** – The COVID 19 led to 15 million deaths and essential health services in 92% of countries got disrupted (2021), leading to reduction in global life expectancy and prevalence of anxiety and depression. Thus, ensuring healthy lives and holistic well-being at all ages is a critical goal.

4. **Quality Education**–Between 2020-21, over half of in-person instruction at schools were missed by 147 million children, and it is estimated that 24 million learners worldwide, from pre-primary to university level, may never return to school. At the same time, many countries have started improving classroom infrastructure as 25% primary schools still lack electricity, drinking water and basic sanitation, and around 50% lack computers and internet access. In addition, skill and competence development is a driver for industry readiness, improved economy and social upliftment, making quality education a major goal.
5. **Gender Equality**–While there was a rise in women employment in 2019, the 45% global employment loss in 2020 set back the equal representation pace. It was also revealed through a 15 year survey, that 1 in every 4 women, accounting to 641 million women, were subject to violence at least once in their lifetime. Therefore, exalting women is decisive towards a better and inclusive future generation.
6. **Clean water and sanitation** – World's water ecosystems are degrading at alarming rates, with over 85% of wetlands being lost, while 733+ million people live in countries with high and critical levels of water stress. This goal aims to ensure availability and sustainable management of water and sanitation for all, as 4 times increase in requirement is estimated by 2030.
7. **Affordable and clean energy**–While progress in energy efficiency is underway, and total renewable energy consumption has increased, the annual energy-intensity rate needs to go up from 1.9% presently to 3.2% by 2030. The key hurdles are; slowdown in electrification due to the challenge of reach, the use of inefficient and polluting cooking systems, and decline in financial flows to develop countries for renewables, making affordable and clean energy a priority.
8. **Decent work and Economic growth** – While global unemployment plummeted, child labour worldwide continues to account to 1 in 10 children. Presently annual growth rate of global real GDP per capita got affected by rising inflation, supply-chain disruption, policy uncertainties, etc. Thus, full, and productive employment is crucial for decent work and sustainable economic growth.
9. **Industry, innovation and infrastructure** – While global manufacturing has bounced back and is on a steady rise, over the COVID crisis it was noted that high-tech industries are more resilient than lower-tech counterparts, with small scale industries lacking financial support and loss in manufacturing jobs. Therefore, building resilient infrastructure, promote inclusive and sustainable industrialisation and foster innovation is key.
10. **Reduced inequalities**–The pandemic caused a first rise in between-country income inequality, and global refugee figures hit a record high. In addition, discrimination on at least one of the grounds prohibited under international Human Rights Law is still faced by 1 in 5 people, making reducing inequality a major global goal.
11. **Sustainable Cities and Communities**–As rapid urbanisation occurs, issues of; polluted air breathed in by 99% of urban population, municipal solid waste of which only 55% is managed, and 1 billion slum dwellers indicate the need to make cities and human settlements inclusive, resilient, and sustainable.
12. **Responsible Consumption and production** – With reliance on natural resources increasing over 65% globally from 2000-2019 and large amounts of food is either lost between harvesting and reaching markets, or wasted at the consumer level, making sustainable consumption a pivotal goal.

13. **Climate Action** – Droughts estimated to displace 700 million, extreme weather, and other natural disasters estimated to increase by 40% by 2030, climate catastrophes are surmounting. 70-90% of coral reefs have diminished and CO_2 emissions have gone up to its highest in 2021, while climate finance has dropped with a shortfall of 100 billion dollars in 2019; showcasing the grim state of affairs and the urgency for climate action as a global goal.
14. **Life below Water**–Oceans are our planet's largest ecosystem and is endangered due to ocean warming, eutrophication, acidification, over-fishing, and plastic pollution; with over 17+ million metric tons of plastics choking the ocean and acidification hindering the ocean's capability to absorb CO_2 emissions to moderate climate change. Conservation and sustainable use of marine resources is a necessity.
15. **Life on Land** – Almost 90% of global deforestation is due to cropland expansion and livestock grazing, and around 40,000 species are at the risk of extinction. Protecting, restoring, and promoting sustainable use of terrestrial ecosystems, managing forests sustainably, combating desertification and address land degradation and biodiversity is our moral duty.
16. **Peace, Justice and strong Institutions** – With a quarter of the world's population living in conflict-affected countries and corruption and bribery rampant, providing justice for all, and building effective, accountable, and inclusive institutions at all levels is imperative for peaceful societies.
17. **Partnerships for the Goals**–Post pandemic, with rising debt burdens threatening developing countries and global Official Development Assistance (ODA) declining for SDG data, it is vital to strengthen the means of implementation and revitalise the global partnerships for sustainable development.

Each SDG has a number of **Targets**, each measurable by indicators; for example, Target 1.1 '*By 2030, eradicate extreme poverty for all people everywhere, currently measured as people living on less that $1.25 per day*' can be measured through the **indicator** 1.1.1 '*Proportion of population living below the international poverty line by sex, age, employment status and geographical location (urban/rural)*'. India, as the fifth largest economy and the second largest in population in the world, plays a crucial role in the achievement of Agenda 2030 and the future. The ***SDG India Index***, an aggregate measure presented in a comprehensible, interactive dashboard (2020-21; http://sdgindiaindex.niti.gov.in), for 16 goals with 115 quantitative indicators and a qualitative assessment of SDG 17, tracks and monitors progress of all states and union territories.

Fig.1.8: Sustainable Development Goals *(source : www.sdgs.un.org)*

As the world moves towards sustainable development, the role of civil engineering becomes pivotal, needing to shoulder the responsibilities of designing and developing sustainable solutions.

1.4 EVALUATING FUTURE REQUIREMENTS

Brundtland's Commission (1987), in its definition of Sustainable Development, stressed on the *"future generations" and the ability to "meet their needs".* These needs and whether or not the ability to attain them by future generations can be assessed by sustainability indicators attributed to each SDG and its respective targets and are measured with various monitoring applications that capture quantitative data, further collated and represented in Indices to offer a holistic view on sustainable development.

1.4.1 Sustainability Indicators

Sustainability indicators are measurable aspects of the three dimensions of sustainability – social, environmental, and economic, and are essential for monitoring and calibrating the performance and quality of the sustainability goals. In addition, these help in decision-making by providing aggregated information to incorporate physical and social science into actionable items and help in predicting early warnings to prevent setbacks.

Division of Sustainable Development (DSD) and Statistics, under the UN Department of Economic and Social Affairs drafted the first set of indicators, which was later collated with methodology sheets for each indicator into a single publication known as the 'blue book' (1996). The same was revised in 2001 having 58 indicators, classified into themes, such as, poverty, governance, health, education, demographics, natural hazards, biodiversity, consumption, and production patterns, etc. Every theme, in turn was categorised into sub themes with core indicators, for example – Poverty is sub-thematised into; '*income poverty*' measured by the indicator 'proportion of population living below national poverty line', and '*income equality*' measured by the indicator 'ratio of share of national income of highest to lowest quintile', etc. Certain sub-themes maybe measured by several indicators, such as, '*Education level*' is indicated by: 'gross intake ratio to last grade of primary education', 'net enrolment rate in primary education', 'adult secondary (tertiary) schooling attainment level'. Interestingly, several indicators have links to more than one theme, such as, 'proportion of population with access to safe drinking water' is primarily applicable to Poverty and Health, but also has secondary linkages to governing water utilities, and in turn, to Governance theme.

1.4.2 Monitoring: Methodology and applications

Monitoring is a major task for assessing the implementation and impact of a strategy for sustainable development. A variety of data and statistics are required for monitoring and to capture this, methodologies to measure the indicator, collect the data accurately and timely, outline the apt unit of measure and method of computation, etc are developed by three key bodies, namely the UNSC, HGL-PCCB, and IAEG-SDGs. The latter has selected custodian agencies, such as, UNEP, WHO, World Bank, FAO, UNESCO, OECD, etc, for periodically collecting and updating data on indicators, across themes, at global, regional, and national levels. **Ten principles for Global monitoring indicators** (UNSDSN, 2015) entail;

1. Limited in number and globally harmonised,
2. Simple, single-variable indicators, with straightforward policy implications,
3. Allow for high frequency monitoring,
4. Consensus-based,

5. Consensus-based, in line with international standards and system-based information,
6. Constructed from well-established data sources,
7. Disaggregated (by sex, age, location, income, spatial–rural/urban)
8. Universal,
9. Mainly outcome-focused,
10. Science-based and forward-looking,
11. A proxy for broader issues or conditions.

Indicators to be measured are categorised with respect to global, national and thematic, and Data in turn, is sourced from census data and *Household surveys* (Demographic and health surveys, Fertility and Family Surveys, Reproductive health surveys, Labor Force survey, R&D Surveys etc); *Administrative data* (formal waste collection and management data from municipalities, National production, international trade statistics); Civil registrations (birth certificates, death certificates, etc), *Vital statistics* and *health records*; *School-based or citizen-led learning assessments*; as well as from *Geographic Information Systems* (GIS) that uses remote sensing, GPS (global positioning system), aerial photographs, LiDAR (Light Detection and Ranging) and satellite imagery from US Geological Survey/NASA Landsat data, for mapping and surveying.

GIS is an application that uses geographical and spatial data in conjunction with attribute (additional information in tabular form) data to map, analyse and assess indicators. It uses imagery data-type, which includes aerial photos, satellite images, thermal images digital elevation models, scanned maps, land maps, land classification maps, etc. Satellites are used to scan the earth, either through–scanning mirrors that scan quickly, back and forth over an area while taking an image, as used by Landsat or Multispectral scanners, called '*whiskbroom scanning*', or via '*push broom*' scanning where a row of silicon detectors take images as the satellite flies over an area. GIS not only aids in cartography but helps geospatial mapping and analysis of real-world problems to offer insights through visualisations. The '*Global Geodetic Reference Fame*' (UNDESA) allows the precise determination of locations and quantification of changes, useful for various indicators through mapping physical infrastructure, buildings and settlements, population distribution, transport networks, elevation and depths, water and land cover and use, etc.

1.4.3 Human Development Index and Ecological Footprint

While SDGs are a tool for addressing developmental progress, there are other indices that also help measure the impact of human development.

One such, is the **Human Development Index (HDI)** that offers an "*alternative, single number measure of capturing progress in three basic dimensions of human development: Health (life expectancy at birth), Education (expected and mean years of schooling) and Living standards (Gross National Income per capita)*" as defined by the Human Development Report (UNDP). This index was further refined and reintroduced in 2010 as '*Inequality adjusted HDI*' (IHDI) accounting for the inequalities between the various nations, noting that HDI maybe perceived as a potential while IHDI reflects the reality. Another important indicator is the **Ecological Footprint (EF)** that measures the human–a person or a group, demand on natural capital. It estimates the productive land and sea area needed to support a population in terms of its consumption of renewable resources and absorption of waste generation. It shows whether a country is living within the biocapacity of its own territory or is drawing on the ecological capital of other nations. Both HDI and EF have attracted criticism owing to aggregation as a single number upon summation of different indicators, each having different units, and in turn, missing the finer details that represents the scenario realistically.

The upcoming decades, bound by a common developmental discourse since the industrial revolution, and further today, by digitisation and globalisation, faces the inadvertent need to strive towards Sustainable development and to understand the importance of innovation and social responsibility, particularly through Civil engineering in shaping and impacting the world.

UNIT SUMMARY

The Unit covers, in brief, the developmental history from ancient to pre-industrial era, giving an overview of the pockets of civilisations across the world and their prowess in engineering, architectural and construction capability. The Unit further traces the oncoming of the second agriculture revolution and how it prompted the first industrial revolution, followed by decisive events and occurrences termed as further revolutions – green revolution, digital revolution, etc., and culminates the discussion with the impact of the above on the natural environment and social welfare, leading to the importance and need for sustainable development, and how present measures have been instituted to achieve the same.

EXERCISES

I. Multiple Choice Questions

Q.1 What is the system of transporting water from aquifers and wells to the surface through underground aqueducts, found in Persia and Mesopotamia?

a. Baoli b. Qanat
c. Reservoir d. Hammam

Q.2 When did the 'Green Revolution' take place?

a. late 19th century b. early 20th century
c. 10,000 BC d. mid 20th century

Q.3 Who coined the term "Gilded Age"?

a. Mark Twain b. FD Roosevelt
c. Henry Ford d. Earnest Hemingway

Q.4 Under who's leadership was C-DOT (Centre for Development of Telematics) established in India?

a. JRD Tata b. Manmohan Singh
c. Sam Pitroda d. Sundar Pichai

Q.5 In which year was the '2030 Agenda for Sustainable Development' adopted?

a. 2020 b. 2015
c. 1987 d. 2000

Answers of Multiple Choice Questions				
1.1 **b**	1.2 **d**	1.3 **a**	1.4 **c**	1.5 **b**

II. Short and Long Answer Type Questions

Q.6 What is Sustainable Development? What are the SDGs?

Q.7 What are the different modes of monitoring SDGs? Enlist the Ten principles of Global monitoring indicators as per UNSDSN (2015)?

Q.8 Define Human Development Index (HDI) and Ecological Footprint? What are the shortfalls of these indices?

KNOW MORE

Civil Engineering during the Industrial Revolution in Britain

https://www.designingbuildings.co.uk/wiki/Civil_Engineering_during_the_Industrial_Revolution_in_Britain

India's IT Revolution

https://www.sritechnocrat.com/blog-story-of-india-it-revolution.php

Sustainable Development Goals and Indicators

https://resources.unsdsn.org/indicators-and-a-monitoring-framework-for-sustainable-development-goals- launching-a-data-revolution-for-the-sdgs

https://undesa.maps.arcgis.com/apps/Cascade/index.html?appid=4741ad51ff7a463d833d18cbcec29fff

3D printing technology and application research in India

Rahul, A. V., Santhanam, M., Meena, H., & Ghani, Z. (2019). Mechanical characterization of 3D printable concrete. *Construction and Building Materials*, *227*, 116710.

Dey, D., & Panda, B. (2023). An experimental study of thermal performance of 3D printed concrete slabs. *Materials Letters*, *330*, 133273.

REFERENCES AND SUGGESTED READINGS

Books and Publications

- [Book] Fletcher, B. (1896). *A History of Architecture on the Comparative Method*. Batsford. [Book] Stearns, P. N. (2020). *The industrial revolution in world history*. Routledge.
- [UN Report] Brundtland, G.H. (1987) *Our Common Future: Report of the World Commission on Environment and Development*. Geneva, UN-Dokument A/42/427.
- Blasi, C., Coïsson, E., & Iori, I. (2008). *The fractures of the French Panthéon: Survey and structural analysis*. Engineering fracture mechanics, 75(3-4), 379-388.
- Manian, M.A.; Khoshbakht, K.; Mahmoudi, H.; Liaghati, H. *Dynamic Conservation in Risk Society: A Case Study of COVID-19 Pandemic Risk in Kashan Qanat Irrigated Agriculture*. Front. Public Health 2022, 10, 882943.

Web sources

- https://www.un.org/sustainabledevelopment/sustainable-development-goals/ https://sdgs.un.org/goals
- https://resources.unsdsn.org/indicators-and-a-monitoring-framework-for-sustainable-development- goals-launching-a-data-revolution-for-the-sdgs
- https://undesa.maps.arcgis.com/apps/Cascade/index.html?appid=4741ad51ff7a463d833d18cbcec29fff

02 IMPORTANCE OF CIVIL ENGINEERING

UNIT SPECIFICS

Through this unit we have discussed the following aspects:

- ***The importance of Civil Engineering in shaping and impacting the world***
  - *Responsibilities and Role in reaching SDGs*
  - *Direct and indirect impact with respect to SDGs*
- ***The ancient Marvels and modern Wonders in the field of Civil Engineering***
  - *Ancient Civilizations [4000 – 900 BCE]*
  - *Classical Period [1000 BCE – 1000 CE]*
  - *Renaissance & Age of Enlightenment [1400 – 1750 CE]*
  - *Modernism & Industrial Era [1750–1950 CE]*
  - *Contemporary style & Digital Era [1950–present]*
- ***Future Vision for Civil Engineering***

Besides giving a large number of multiple choice questions as well as questions of short and long answer types marked in two categories following lower and higher order of Bloom's taxonomy, a list of references and suggested readings are given in the unit so that one can go through them for practice.

There is a "Know More" section, which has been carefully designed so that the supplementary information provided in this part becomes beneficial for the users of the book. It is important to note that for getting more information on various topics of interest some QR codes have been provided which can be scanned for relevant supportive knowledge This section mainly highlights applications of the subject matter for our day-to-day real life or/and industrial applications on variety of aspects, case study related to environmental, sustainability, social and ethical issues whichever applicable, and finally inquisitiveness and curiosity topics of the unit.

RATIONALE

This unit establishes the importance of the profession and practice of Civil engineering in the context of Sustainable development and future trends and needs. It further emphasizes the roles and responsibility of the civil engineer.

UNIT OUTCOMES

List of outcomes of this unit is as follows:

U2-O1: Knowledge on the role, impact/relevance, and importance of Civil Engineering
U2-O2: Knowledge on the outstanding Civil Engineering feats of ancient and modern times
U2-O3: Knowledge on the future trends in the field

Unit-2 Outcomes	EXPECTED MAPPING WITH COURSE OUTCOMES (1- Weak Correlation; 2- Medium correlation; 3- Strong Correlation)						
	CO-1	CO-2	CO-3	CO-4	CO-5	CO-6	CO-7
U2-01	3		2	3	1	1	3
U2-02	3	2	2	1	2	1	
U2-03	2		1	2			3

The Civil Engineering discipline encompasses various specialisations, like, Construction Engineering, Structural Engineering, Geotechnical Engineering, Transportation Engineering, Environmental Engineering, Materials Engineering, Marine Engineering, Irrigation Engineering, Highway Engineering, Bridge Engineering, Hydraulic Engineering, etc. The built- environment, comprising of buildings, bridges, railways and roads, sewers and dams, power pants and transmission towers and lines, tunnels, canals, and waterways, are all results of these sub-domains. As discussed earlier, the built-environment, along with the inhabitants and social ecosystem it supports, has profound impact on the world and is a critical determinant of sustainable development. Hence, not only is civil engineering of importance for present needs of shelter, sanitation, transportation, and connectivity, but plays a significant role in addressing the issues and concerns arising due to the unmitigated urbanisation and economic development.

In the following Sections, a brief discussion on the importance of the discipline in shaping and impacting the world is followed by tracing the evolution and development of civil engineering practice through examples across time – from ancient era to present society. These ancient Marvels and modern Wonders reflect the society in which it thrived, with respect to the events and context discussed earlier in Unit 1, and paves way for the innovations of the future.

2.1 CIVIL ENGINEERING: SHAPING AND IMPACTING THE WORLD

The civil engineering market is expected to grow 25% each year (2022) up to 11.7 trillion US Dollars by 2025 (Global Market Insights, Inc.) due to the ever-increasing demand for infrastructure. In turn, increasing the need of competent engineers possessing sound knowledge of engineering fundamentals, and software and technical proficiency. In addition, a Civil engineer must hone skills as identified as the **Top Skills for 2025**, such critical thinking and creative problem-solving skills, the ability to communicate and collaborate effectively, coupled with strong leadership required for project management (World Economic Forum). Beyond the core activities of designing, i.e., conceptualise novel solutions, develop layouts and perform design calculations, and On-site supervision, civil engineers are **responsible** for;

- Oversee or perform soil tests, surveying operations, materials test, etc., and Analyse survey reports, maps, and other data to plan projects.
- Design and communicate ideas upon interacting with all stakeholders and identify possible design improvements, as well as incorporate Building codes, standards and guidelines into the design as required.
- Develop Project scope and timeline, and manage and monitor each stage of project,

- Assess environmental impact and risks, and ensure job site meets all legal guidelines and health and safety rules.
- Assist with staging, testing, and shipping of equipment prior to deployment.
- Prepare and present technical reports, analysis reports, cost estimates, Bill of Materials (BOM) and Bill of Quantities (BOQ), environmental impact statements.
- Submit permit applications to local or national agencies, and mitigate conflict.

Civil Engineering plays a **significant role** in offering methodologies, approaches, assessments of risks and impacts, as well as technologies and tools to assist decision-makers and technicians to achieve the Sustainable Development Goals (SDGs), in the following ways;

i. ***Planning measures to mitigate and adapt*** to climate change, extreme weather events, earthquakes, droughts, floods, and other natural disasters.

ii. ***Developing efficient and sustainable strategies*** for resource utilization while minimizing environmental impact and addressing unequal distributions.

iii. ***Enhancing the safety of structures and infrastructures*** against exceptional loads and deterioration over their lifecycle.

iv. ***Implementing a comprehensive risk management*** approach and appropriate technologies to reduce pollution and environmental degradation, thereby reducing vulnerability.

v. ***Establishing safe drinking water and sanitation systems*** to safeguard human health.

A huge responsibility of achieving the SDGs lies on the shoulders of Civil engineers and the need for creative and sustainable solutions is the need of the hour. Most of the 17 SDGs are in some way connected to the discipline of Civil engineering, directly or indirectly, as illustrated below, highlighting the importance of the discipline.

- The **direct impact** that Civil engineering interventions can have, are on *SDGs 6 – Clean water and sanitation, 7 – affordable and clean energy, 9 – Industry, innovation and infrastructure, 11 – sustainable cities and communities, and 12 – Responsible consumption and production.*
- Under *SDG 1 – Eradicate Poverty*, its associated Target 1.4 focuses on "*access to basic services, ownership and control over land and other forms of property, inheritance, natural resources*", and civil engineers play an active role in planning land use.
- Under *SDG 2 – Zero Hunger*, Target 2.3 outlines "*By 2030, double the agricultural productivity*" and Target 2.4 stresses on "*By 2030, ensure sustainable food production systems and implement resilient agricultural practices*", which requires civil engineering interventions in form of irrigation, water and waste management, transportation, etc.
- SDGs 3 – *Health and Well-being, 4 – Quality Education, 8 – Decent work and economic growth, 13 – Climate Action, 14- Life below water, and 15 – Life on Land* can all be improved by Civil engineering, through infrastructural solutions with reduced environmental impacts that are socio-economically respondent.
- Further, civil engineering projects and practice can become a potent medium to create awareness and statements, inspire change and imbibe best practices in the society towards *SDGs 5–Gender Inequality, 6 – Reduced Inequalities, 16 – Peace, justice and strong institutions.*

2.2 ANCIENT WONDERS AND MODERN MARVELS

The prowess of a civilisation can be contemplated through the tangible and intangible cultural heritage, and the feats of design, engineering and construction of the ancient world indicates the advancement and creativity of these long-gone people. While events and aspirations dictated the progress of the civil engineering practice, the ancient wonders and modern marvels showcase the importance and impact of the discipline on mankind.

2.1.1 Ancient Civilizations [4000 – 900 BCE]

One of the first, most formidable, examples of civil engineering and urban planning maybe noted in the Indus Valley, with the city of Harappa being built during the Bronze Age, in the 4th millennium BC. Sanitation and wastewater systems were uncovered dating back to 2550BCE in the cities of Mohenjo-daro, Harappa and Lothal, and well planned urban housing made of strong *leeves* or earthen walls, with private toilets and drainage, networks of reservoirs and canals for irrigation, granaries with air ducts raised on high platforms, and designated public baths (refer Fig. 1), are some of the remnants of civil engineering of these flourishing cultures in the valley.

Fig. 2.1: Great Bath at Mohenjo-Daro

(source : https://www.harappa.com/sites/default/files/slides/bath-indus.jpg)

In parallel, in ancient Mesopotamia (Iraq) along the rivers of Tigris and Euphratyes, the stepped pyramidal structure referred as **Ziggurat, at Ur** dates to 4000 BCE, with the 'White Temple' on top of it being built in circa 3500 BCE. Underwater channel systems, called '*qanats*', were constructed to route water from aquifers and wells to the surface for consumption and irrigation.

Civil engineering and architecture were at a peak in ancient Egypt. The **stepped pyramid** dedicated to King Djoser, of the 3rd Dynasty, located at Saqqara Necropolis in Egypt, was erected around 2700 BCE. Designed and build by architect-turned-God Imhotep, documents reveal his profound contribution of using shaped stones to create mammoth monuments with simple tools and mathematics. Shortly after, between 2575–c. 2465 BCE, Pharaoh Khufu of the 4th Dynasty, commissioned the building of the **three principal pyramids of Giza**–Khufu or the Great Pyramid (originally 481 feet); Khafre (471 feet); and Menkaure (213 feet), on the west bank of the Nile. These two sites, along with the ancient ruins of Memphis, are collectively recognised as a UNESCO World Heritage site.

Whilst in Europe, the **Stonehenge**, United Kingdom, was being completed approximately around 2,400 BCE. It comprises of two rings of standing stones with horizontal lintels on connecting them, arranged in a ditch with earthen embankments, with a trilithon – a structure of two large vertical stones with a horizontal resting across them, at the centre. The outer ring has vertical sarsen standing stone measuring 13 feet high, 7 feet wide and weighing roughly 25 tonnes, and the inner ring has smaller blue stones. The Stonehenge in Amesbury, Silbury Hill – a manmade earthen mound with pits and tunnels, and other henges at Avesbury, all are recognised as **UNESCO World Heritage sites.**

2.2.2 Classical Period [1000 BCE – 1000 CE]

What followed was an era of superlative design and construction, characterised by their architectural styles, hand in hand with the rise and fall of empires. From 1st millennium BC to 1st millennium AD, the Greco-Roman style dominated and was termed as 'Classical'. The Greek, and later the Roman, primarily military-oriented, focussed on expansion of their empires and hence, heavily invested on development of infrastructure. **Cities** emerged, such as, the capital Rome in Italy; Persepolis in Greece; Marseille in France; ports at Nucratis and later Alexandria, named after Alexander the Great, in Egypt; and Antioch in modern day Turkiye, once the seat of both the Byzantime and Roman empire. The Roman's championed road building and the first known roadway, the **Appian Way** or the 'queen of the roads', was constructed in 312BCE, connecting Rome with its allies in Capua. The Romans were also the first civilization of built permanent bridges or *Ponte,* as it not only played a strategic role in connecting the vast ends of the ever-growing empire traversing various rivers, but also was perfected in design to behave as aqueducts to carry water. **Pons Amelius**, the oldest stone bridge; Ponte Milvio, the second bridge; and Pons Fabricius, the oldest bridge still standing, all plied over the river Tiber, while the **Pont du Gard,** the tallest Roman bridge, carried water over 50km across southern France to the colony of Nimes.

Temples, urban settlements, such as cities and ports, and Universities were constructed, offering a broader picture of the priorities and wealth of the empires of the time. However, it is interesting to note that there was a strong underlying religious belief system in the Greco-Roman Pantheon and was a centric theme of construction. The **Temple of Artemis** at Ephesus, the **Statue of Zeus** at Olympia, **the Collosus** of Rhodes, **the Pharos** (lighthouse) at Alexandria, and the **Mausoleums at Halicarnassus**, all built at the time, are recognised as amongst the Seven Wonders of the Ancient World (other two being, the Great Pyramids of Giza and the Hanging Gardens of Babylon). Other notable religious structures built at the time are, the **Parthenon** dedicated to Goddess Athena at the Acropolis, a dedicated site atop Athens; **Temple of Jupiter Optimus Maximus** in Rome and those by Herod in the Judea, modern day Jerusalem.

In parallel, between the 10th–5th century BCE, one of the first universities of the world was established at **Takshashila**, near the bank of the Indus River, India. It was a centre of great learning under the Indo-Greeks, with many scholars from all over the world. Chanakya, the noted scholar and Prime Minister of emperor Chadragupta Maurya, was a key figure at the university.

Further north, under the vision of the first emperor of united China, of the Qin dynasty, several existing piecemeal defensive wall structures were connected in the 3rd century BCE to develop a singular system of fortification, consisting of river dikes, bulwarks, and natural terrain, extending form the eastern Hebei province to the Gulf of Chihli. The early **'Great Wall'** was built with rammed earth, stones and wood, and later fortified with bricks and tiles with lime mortar, having passageways blocked by wooden gates. The construction of the Wall flourished under the Ming dynasty and continued till 17th century CE.

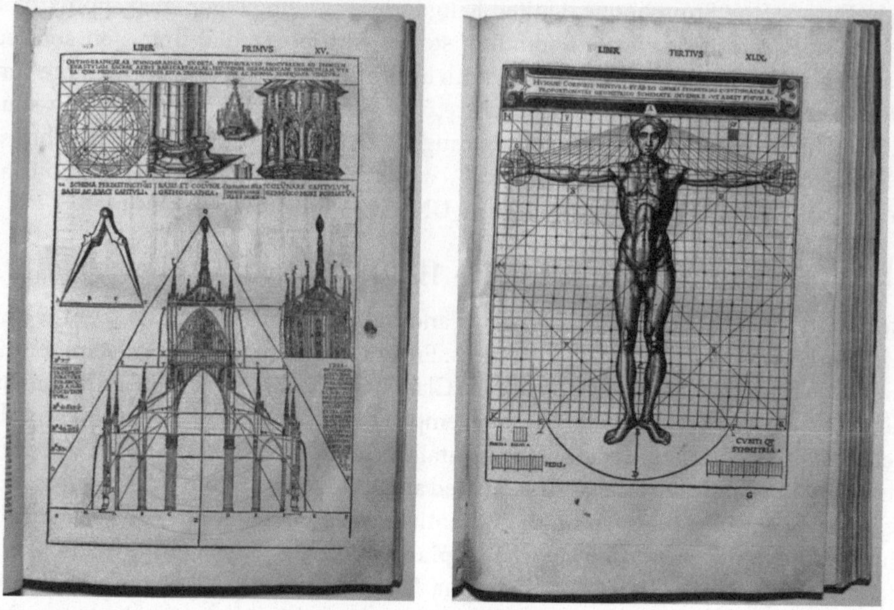

Fig. 2.2: Pages from De Architectura [1475-1543], first vernacular edition, translated by Cesare di Lorenzo Cesariano

(source : https://www.bada.org/object/first-vernacular-edition-de-architectura-vitruvius)

The development and eventual documentation of the **Classical style** can be traced to the earliest temples in Greece, such as that at Samos, built in timber framing. With change in material, from timber to stone, the scale and proportions grew, whilst retaining and refining the design elements and principles, that are still seen prevalent today. This was later composed as a ten-book treatise named, '*De Architectura*' (refer *Fig. 2*), authored by architect and military engineer, Marcus Vitruvius Pollio around 15c. BC, under the patronage of Caesar Augustus. It covered building materials, construction techniques and machines, guidelines on design of temples, civil and domestic buildings, common facilities, such as, pavements, water supply, aqueducts, as well as discussed, scientific and aesthetic principles, such as, geometry, measurement, buoyancy (Archimedes' principle), astronomy for sundials, and '***Orders of architecture***'. In addition, the first book opened with discourse on Town planning, architecture, and civil engineering, and elucidated the qualifications of an architect-engineer. Vitruvius proposed the triad of '*utilitas, firmitas, venustas*' (utility, strength, and beauty), as aesthetic characteristics of a good design and further developed the terms, order, arrangement, fit and proportion, which inspired Leonardo da Vinci to later illustrate the '*Vitruvian Man*'.

The 1st millennium AD saw the loss of Antiquity, as it gave way to political ambitions and philosophical schools of thoughts, led by thinkers such as, Plato and Socrates, leading to a shift from building of religious centres and temples to that of political and social interests, such as, the **Colosseum in Rome** (refer *Fig. 3*) and various triumphal monuments, such as, **Arch of Titus** and **Trajan's column**. But in the second half of the millennium, civil infrastructure continued to be built, such as, **Alcantara bridge** and **baths of Caracalla**; the rise of Christianity once again piqued interest in religious buildings. While several existing Greco-Roman temples and grottoes were converted in Christian churches, the new Roman Christian emperors took upon themselves tasks to build several basilica across their kingdom, from the **Basilica of Maxentius and Catacomb of the Via Latina** in Rome, and the **Papal basilicas** in present day Vatican City; to the *Aula Palatina* or Basilica of Constantine at Trier, Germany, and the **Hagia Sophia** (today known as the Grand Mosque) in Istanbul, Turkiye.

In 5th century CE, another profound institute of education–**Nalanda University,** in present day Bihar, India, came into establishment under the vision of the Gupta empire; and monolithic, rock- carved artistry and construction thrived under the patronage of the Pallava Kingdom in the southern state of Tamil Nadu. The famed Shore Temple, the *rathas* or temple-chariots dedicated to the Pandavas, several *mandapas* or pavilions, and noteworthy rock reliefs, such as '*Arjuna's penance*', at **Malappuram** (or Mahabalipuram), has earned recognition as a UNESCO World Heritage Site. Meanwhile in China, the world's oldest open-spandrel segmental arch-bridge of stone, fondly called Ānjì, meaning 'safe crossing' or **Zhaozhou Qiáo,** was constructed in the Hebei Province between 595-605 CE. In the 8th century CE, the **Seokguram grotto**, as part of Bulguksa Temple complex, in South Korea, was built and is recognised as National treasure 24 and as a UNESCO World Heritage site.

Thence onwards, the story of civil engineering runs hand in hand with architecture, largely responding to the human condition and the social context, with respect to, the visual language and aesthetics, structural layouts and buildings elements, construction techniques and materials, and the artist- engineer's creativity. This led to the development of styles categorised into 'periods' or 'movements'.

Following the Classical period, and before the next grand period of the Renaissance, which translates to 'rebirth', where two shorter styles that were found across Europe, namely;

Romanesque, found mostly across Medieval Europe between 1050-1170, retained several features of the Roman style and is characterized by thick, heavy piers, narrow windows, stained glass, semi-circular arches, and towers, as in the Tower of Pisa; and **Gothic**, or *Opus Francigenum*, meaning "French work", was prevalent in the Late Middle Ages in France, between 900-1300, could be distinguished by pointed arches, high vaulted ceilings, flying buttresses, and vibrant interiors with stained glass windows, gables, colourful tapestries, etc. Noteworthy examples are, the **Cathédrale Notre Dame** in Paris, France, a UNESCO world heritage site and the Milan Cathedral or ***Duomo di Milano***, Milan, Italy.

Fig. 2.3: Different Architectural styles across Italy (*Left*) Colosseum, Rome, Italy

(source : photographed by Author, 2015), (Right) Tower of Pisa, Italy

Fig. 2.4: Examples of Gothic Architecture across Europe : (*Left*) A model of Notre Dame Church exhibited inside the premise, Paris, 2019; (*Right*) A picture of the Duomo di Milano, Milan, Italy, during restoration in 2015

(source: photographed by Author)

2.2.3 Renaissance & Age of Enlightenment [1400 – 1750 CE]

Renaissance, emerged as a response to the over-embellished previous styles drawing inspiration from the Classical in the early 15th century at Florence, Italy, and spread across rest of Italy, France, Germany, Russia, Spain and England. It was characterised by use of arched openings, hemispherical domes, vaulted ceilings, orderly arrangement of columns, and formal landscaped gardens, exemplifying proportion, symmetry, and order. Several artist-architect- engineers, such as, Brunesllleschi, Boticelli, Michelangelo, Alberti, Palladio, Bramante, Bruni and Leonardo da Vinci, and great men of science and philosophy, such as, Copernicus, Galileo, Francis Bacon, Newton, Keplar, Machiavelli and More, flourished in this period. This age of discovery paved the path to several scientific and technological advancements of today.

Filippo **Brunelleschi**, popularly known as the 'Father of Renaissance', designed the *Basilica di San Lorenzo*, housing the pulpit and tondo works by **Donatella**; and engineered the world's largest brick vaulted dome, measuring 45.5m width and 116m height, for the *Santa Maria del Fiore* or Florence Cathedral. Another important advent of the Early Renaissance was the first theoretical ten-book compilation 'on the Art of Building', *De re aedificatoria*, by Leon Battista **Alberti**. It is also the first printed book on architecture, dated 1485, followed by the first edition print of Vitruvius' *De Architectura* (refer Fig. 2).

In the following High Renaissance period, Bramante, under the patronage of the Duke of Milan, incorporated classical styles in contemporary buildings bringing about a unique character to this period. Two noted examples are, the abbey church of *Santa Maria delle Grazie*, where he introduced the choir and a hemispherical dome with arched classical openings, rooted within an octagonal drum. He also worked on the cathedral of Pavia, the cloister of San Pietro in Montorio, and the *Cortille del Belvedere*, Vatican. In 1506, his design of the rebuilding of St. Peter's Basilica was selected, however, owing to his death the new chief Architect and noted artist, **Michaelangelo**, stayed true to the original proposal of the Greek-cross plan by Bramante. Another master builder of the time, **Andrea Palladio**, titled the "most influential architect of the Renaissance", adopted the Classical elements in a less grandiose manner and developed from it unique features, such as, the Palladian Arch–a scaled triumphal arch with square-topped opening for windows, the symmetric domed hall with identical four facades with orders supporting triangular pediments, as in the Villa Capra, famously called 'La Rotonda'. He also authored the "*I Quattro Libri dell'Architettura*" or Four Books on Architecture, published in 1570.

During the Late Renaissance, also called **Mannerism**, the trend continued where classical elements were reinterpreted as is seen by Michelangelo in the **St. Peter's Basilica.** The cupola of the basilica was designed as two masonry shells, one sitting into another, held by ribs which supported a massive roof lantern, and the exterior boasted a 'giant order' (see Fig. 5), defining the external square, held together by a wide cornice – creating a timeless masterpiece.

Following Renaissance, two styles emerged – the **Baroque**, and later, more flamboyant, **Roccoco**, between 1600-1755 as a sharp contrast to Renaissance. With an attempt to apply certain classical elements to regional architecture, without regard for the fundamentals of composition characterising Classical, these two styles are characterised by geometrically exaggerated, ornate and colourful structures, such as the **Palace of Versailles in France**, and **St. Paul's Cathedral in London**, and were seen in conjunction with the Counter Reformation movement.

Fig. 2.5: St. Peter's Square, Vatican, boasting Michelangelo's 'giant order'
(source : photograph by Author, 2015)

2.2.4 Modernism & Industrial Era [1750–1950 CE]

Civil engineering remained at the helm during the **First Industrial Revolution**, with industries springing up and connectivity, transportation and sanitation became a dire requirement. Several bridges, canals and sewers were commissioned to carry effluents, goods and people, to and fro from industries. Noteworthy designs include *Bridgewater canal*, built by engineer Brindley in 1761, having several tunnels allowing direct travel to coal mines; and the world's first cast-iron, arch bridge over the River Severn at Coalbrookedale in 1776, where Pritchard used the brittle metal to construct a semi-circular structure that transfers load onto abutments, while held in compression to counter cracks from propagating – a first of its kind.

In 1771, John Smeaton the first self-proclaimed civil engineer, formed the **Smeatonian Society of Civil Engineers**. He constructed the Eddystone Lighthouse, off Cornwall, England, and developed 'hydraulic lime' and techniques of dovetail joints and dowels for securing blocks.

Road maintenance also became a priority in the 18th century, with the establishment of turnpikes trusts and toll roads. **Design and construction details of roads** saw significant improvement, with new methods proposed by Pierre Tresaguet in 1764 and later, improved by his mentee, John Metcalf who proposed convex shaped surface to allow rainwater to drain. Thomas Telford suggested small stones on the rock foundation further covered with a mixture of stone and gravel, having elevated pavements on either sides, known as 'Telford pitching', and John McAdam developed the '**macadamisation**' where soil foundation is layed with aggregate layers of angular stones and gravel. This design was later patented as the 'Tar Mc Adam road' by civil engineer, Hooley 1901, as is used today.

Soon railways spread rapidly and took over as the preferred mode of heavy goods transportation, with engineer Stephenson designing the **first steam locomotive** in 1825, and later improving it under the *Liverpoool and Manchester railway* (L&MR) in 1830.

A prolific civil engineer of the time, **Isambard Brunel,** pushed the boundaries of engineering and expanded his role as engineer-architect and designer. By 1833, the *Great Western Railway* (GWR) was commissioned under the able guidance of who had assisted on the *Thames Tunnel* project which has been incorporated today as part of the London Underground. Brunel, also has to his credit, design of several remarkable bridges, with the *Cliffton Suspension bridge* over Avon spanning 702 m–the longest at the time, as well as for, the world's longest steamship of the time, '*Great Western*', constructed of wood and reinforced with iron diagonals and bolts, with improved surface condenser incorporated. He continued his experiments and further designed the first modern ship, *Great Britain*, which was propeller-driven and iron-hulled. The *Great Eastern*, his third transatlantic ship design, was the most technologically-advanced and luxurious ship of the time and is famed as one of the seven wonders of the Industrial Revolution. However, it aided the laying of oceanic telegraph cables rather than plying passengers, and proved to be pivotal in connecting America and Europe. He also designed the '*Three bridges*', London, that allowed the routes of Great Western and Brentford Railway, Grand Junction canal and Windmill Lane; and the Renkioi Hospital, a first of its kind hospital in 1845, comprised of pre-fabricated wood and canvas that could be shipped and assembled on-site, far away in Tukiye, where the Crimean War was underway.

During the **Second Industrial Revolution**, when the spotlight was on the new nation of United States of America (USA), the **Neoclassical style,** as the name suggests, a revival of Classical style, was adopted to communicate a grand and powerful aesthetic in Gilded Age. Characterised by triangular pediments, free-standing columns, balustraded balconies, pronounced cornices, and symmetry, the White House and overall, the Capitol complex in Washington DC, it is reminiscent of Palladian architecture. Neoclassicalism also spread to England, France and Russia, with some of its outstanding advocates being, Robert Adam, John Soane and Claude-Nicolas Ledoux.

Modern design as understood today, epitomized by metal construction, first sprouted its presence with the **first two exhibition pavilions of the World Fairs** or the "*Exposition Universelle*". The two were, the **Crystal Palace** in Hyde Park, London, a cast-iron and plate-glass structure housing 92,000 sqm exhibition space designed by Joseph Paxton in 1851; and the **Eiffel Tower,** Paris, locally referred to as "*La dame de fer*" (Iron Lady), a wrought-iron lattice tower designed by architect, Stephen Stauvestre, along with structural engineers, Maurice Koechlin and Emile Nouguier. With further industrialisation supporting mass production of steel and glass, these materials became vastly employed as they imbibed a sense of never-before appeal, and further motivated futuristic styles.

Fig. 2.6: The Great Exhibition pavilions–(*Left*) Crystal Palace, London, (*Right*) Eiffel Tower, Paris

Fig. 2.7: Examples of Art Nouveau (*Left*) Sagrada Familia, Barcelona, and Art Deco style (*Right*) Chrysler Building, Chicago

Between 1890-1910, aversive to historicism and enthusiastic of modern life, the unique style of **Art Nouveau** developed characterised by organic lines and sinuous forms, achievable by steel frames and exaggerated by glass panes. The style was called *Jugendstil* in Germany, *Sezessionstil* in Austria, *Stile Floreale* (or Stile Liberty) in Italy, and *Modernismo* (or Modernista) in Spain; and was employed in building design by architects, Louis Sullivan across Chicago and further accentuated by artist-architect, Antoni Gaudi in Spain. Sullivan advocated the development of original forms and ornamentation, and is identified as **'Father of skyscrapers'**, exemplified in his design of the Wainwright Building in Chicago. In his firm worked a young civil engineer and aspiring architect, Frank Llyod Wright, who later was called the **'Father of modern architecture'** and like his mentor, developed a style unique to America – the Prairie style, illustrated through large cantilevers and use of glass held with metal, like the Robie house or 'Fallingwater' – the most famous house in the world.

It was followed by **Art Deco style**, also referred to as style modern, which exhibited an affinity with machines as a reflection of modernity. While a short-lived movement between the 1920's and 30's, it had an indelible mark on the skyline of the emerging nation of USA, with iconic structures, such as, the Empire State Building, Rockefeller Center and Chrysler Building. This style was characterised by stepped gables, sculptured panels, ornate geometry, and cubic forms, along with use of unorthodox materials like, exposed steel and aluminium, decorative glass, ceramic and even, stucco and terracotta. In contrast, Europe saw the development of the '**International style**' characterised by rectangular forms, cantilevered projections, flat roof, ribbon windows, curtain glass, asymmetric facades and lack of ornamentation, as practiced by Le Corbusier in France, and Walter Gropius and Mies van der Rohe in Germany.

Various other styles such as, *Futurism, Constructivism, Brutalism, De Stijl, and Bauhaus*, all encompassed under the term **'Modernism'**, prevalent during the early 20th century, were grounded on the principle of '*Form follows Function*' as stated by Louis Sullivan. Fuelled by the abundant availability of mass-produced modern materials; they were a stark deviation from the traditional styles, characterised by functionalism, lack of ornamentation, and rational use of modern materials true to their nature, with a keenness for structural innovation. Philip Johnson's Glass House is an exquisite example.

2.2.5 Contemporary style & Digital Era [1950–present]

Post-World War II, the international style flourished in the US for commercial buildings, as evident in the Seagram Building designed by Mies van der Rohe, a champion of minimalism famous for his aphorism "*less is more*".

But soon, in the 1960's, there was an antagonism towards the bleakness of modernism, and newer trends, such as, the introduction of the principles of **Beaux Arts style** by Louis Kahn and Eero Saarinen; and the need to focus on placemaking, with the local conditions and contexts through incorporation of vernacular by Robert Venturi, who published '*Complexity and Contradiction in Architecture*', gave rise to the **Post-Modernism.** It challenged its predecessor with asymmetry, ornamentation, historical details and familiar motifs, and was characterised by eclectic inspirations and kitsch aesthetics, with focus more on form over function. Some of the notable examples are, the Lotus Temple in New Delhi, the Sydney Opera House, and the Guggenheim Museum, Bilbao, by pioneer Frank O'Gehry.

The concept of modularity was also catching on, with Archigram collective proposing a mega-scale modular concept called Plug-in City in 1964 and Shafie Moshdie designing the Habitat 67, a modular experimental housing. In Japan, the '***Metabolism' style*** developed, focusing on modularity, flexibility and interchangeable units, as exemplified by the Nagakin Capsule Tower designed by Kisha Kurokawa in 1972.

Fig. 2.8: Examples of Modernism–(*Left*) Frank Llyod Wright's *Fallingwater*, Pennsylvania, (*Right*) Philip Johnson's *The Glass House*, Connecticut

Fig. 2.9: Examples of Contemporary Style–(*Top Left*) Frank O'Gehry's Guggenheim Museum, Bilbao, (*Top Right*) Nagakin Capsule Tower Kisho Kurokawa's *Nagakin Capsule Tower*, Tokyo, (*Bottom Left*) Mies van der Rohe's *Seagram Building*, New York City, (*Bottom Right*) Renzo Piano and Richard Roger's *Centre de Pompiduo*, Paris

Meanwhile in India, between 1950's-70's, the young nation also felt the ripples of these movements, with the vision of setting up a model city – Chandigarh. The task of designing a modern yet culturally sensitive aesthetic, Le Corbusier devised the city's masterplan by applying the concept of *Unité d'Habitation*. With the Capitol Complex at its heart, clean geometry and concrete facades were interspersed with motifs and symbology, and compositions of colour. And in 1961, the year of completion of the construction of the Capital Complex – a UNESCO World Heritage site, another grand project was in motion. Louis Kahn was commissioned the tasks of designing the IIM Ahmedabad (1962-74), a vision of Dr. Vikram Sarabhai and Kasturbahi Lalbhai, to develop professionals for the country's growing industrial progress.

The introduction and widespread adoption of the use of CAD software between the 1960's – 1990's led to the widening of architectural styles beyond post-modernism to high-tech or **Structural expressionism**, which aimed at showcasing the underlying function-structure of the buildings, from exterior to the interior, through the use of advanced technology and materials. Aluminium, steel, glass and concrete offered a sense of grandiose and honesty, with open plans allowing reconfigurable spaces, large overhangs and lack of load bearing walls. The World Trade Centre (1971) in New York by Yamasaki, the Centre Pompidou (1977) in Paris by Renzo Piano and Richard Rogers, and the Burj al Arab (1991) by Tom Wright are some famous examples. Improved design tools furthered modular construction and led to the development of the **Klip House concept**, designed by Interloop between 1997-2001, where modules could be snapped together to build a unit. Presently, the world's tallest building, the Burj Khalifa (2010) with the height of 830m to the tip, housing 163 floors, is an epitome of contemporary architecture. The three-leafed structure is an abstraction of the desert flower, Hymenocallis, twirling into a spire as it gains height.

Hand in hand, another trend of the present times, propelled by the development of CAD and other design tools, is **Parametricism.** Characterised by the use of algorithmic equations and computational tools leveraged to generate varied and almost impossible forms which are complex yet fluid, can be traced back to 1997. Earlier famed architects, Frei Otto and Antoni Gaudi are often considered inspirations for this style, personified by *Pritzker Awardee, Zaha Hadid.* The Guangzhou Opera House, at Galaxy Soho in Beijing China, and the Hyder Aliyev Centre in Baku, Azerbaijan, are a few of her exemplary works.

While using parametric design not only enables designers to optimise the planning and improve efficiency of the design, attempts to incorporate sustainability factors is currently being pursued by noted firms, such as, Ai Space Factory, Foster & Partners, etc. The OPPO R&D headquarters, named the **'Infinity Loop',** *designed by Bjarke Ingels Group (BIG)*, China, boasts an elegant form that is self-shaded, thereby, it reduces the energy usage and increases natural light. Another remarkable example is the '*Heart of Yong'an*' by TJAD, China, where algorithmic modelling has been used to create a hyperbolic, single slope, curving roof structure, emulating the surrounding mountains, with rammed earth walls and local blue tile roofing.

Fig. 2.10: Examples of Parametricism–(*Top*) Views of Zaha Hadid's *Hyder Aliyev Center*, Baku, Azerbaijan, (*Bottom*) of Bjarke Ingels *Group's Infinity Loop*, Hangzhou, China

2.3 FUTURE VISION FOR CIVIL ENGINEERING

With demand for design and construction on the rise, the future holds many propositions. The key themes for civil engineering, at the juncture of **design + technology**, will be:

Sustainable Design & construction: Developing ***Green infrastructure*** that is safe, inclusive, energy efficient, and uses materials and techniques that have low environmental impact; using ***Nanotechnology and use of nano materials***, like, nano-silica, nano-clays, and iron nanoparticles, incorporated into concrete, and copper nanoparticles used in steel beams; and realising ***Futuristic transportation***, such as self-drive and travel pods through hyperloop will improve quality of life.

Cutting-edge technology: Using ***Augmented and Virtual reality*** to model real world construction challenges, as well as provide an immersive experience to visualise concepts via ***Digital Twins***; and leveraging ***AI, Big data and IoT*** to seamlessly harness data for mundane decisions and support collaborative projects with multiple stakeholders, will improve the quality and time of the design and construction.

The importance and profound impact of Civil engineering on society and the world, is undeniable and beckons that formidable and competent civil engineers forge the future.

UNIT SUMMARY

This Unit walks through an aspiring civil engineer on the importance of the profession, its remarkable lineage of breakthroughs across history and the future scope of the profession. Drawing from the knowledge shared in Unit 1, this Unit connects to the importance of the profession and practice as custodians of sustainable development and exemplifies the potential impact of civil engineering on achieving each SDG, either directly or indirectly. It further discusses the skills required for a competent civil engineer and the capabilities expected. The next Section elucidates marvels and breakthroughs in the ancient era up till present times, and continues to highlight in brief the major themes for the future.

EXERCISES

I. Multiple Choice Questions

Q.1 Which of the following is not a Civil Engineering specialisation?

a. Geotechnical Engineering b. Hydraulic Engineering
c. Environmental Engineering d. Urban Planning

Q.2 What are the SDGs that are directly impacted by Civil Engineering?

a. SDG 11–Sustainable cities and communities,
b. SDG 6–Clean water and sanitation
c. SDG 16–Peace, justice and strong institutions
d. SDG 9–Industry, innovation and infrastructure

Q.3 Which ancient marvel, recognised as a UNESCO World Heritage site, is the oldest amongst the Seven Wonders of the World?

a. Three principal pyramids of Giza b. Ziggurat with the 'White Temple' at Ur
c. Mohenjo-daro, Harappa and Lothal d. The Stonehenge, United Kingdom

Q.4 Q. 2.4 The Greeks and Roman style of architecture and construction is called?

a. Renaissance b. Ancient
c. Classical d. Gothic

Q.5 Who is the Father of Renaissance?

a. Leonardo da Vinci
b. Andrea Palladio
c. Leon Battista Alberti
d. Filippo Brunelleschi

Answers of Multiple Choice Questions				
2.1 **d**	2.2 **c**	2.3 **a**	2.4 **c**	2.5 **d**

II. Short and Long Answer Type Questions

Q.6 What are the roles and responsibilities of a Civil Engineer?

Q.7 What were the major periods in design and construction? Write briefly on each with an example of civil engineering.

Q.8 Describe the major future trends in Civil Engineering and elaborate which one is set out to be the most impact full in the days to come.

KNOW MORE

About Mohenjo-Daro

https://study.com/learn/lesson/harappa-mohenjo-daro-indus-valley-civilization-history-decline.html

https://youtu.be/KhDY4KJuvc0

About Mesopotamia

https://www.youtube.com/watch?v=EwY-ziBL1Jw

About the Seven Wonders of the Ancient World:

https://youtu.be/1YWPMZ6pHm8

Timeline of Architecture styles:

https://www.worldhistory.org/timeline/architecture/

https://urbandesignlab.in/timeline-of-the-history-of-architecture/

About Renaissance:

https://www.thecollector.com/10-characteristics-of-renaissance-architecture/

http://www.visual-arts-cork.com/architecture/renaissance.htm

Road construction

https://dreamcivil.com/history-of-road-development/#_3_Tresaguet_Construction

History of CAD and Design model making

https://www.creativemechanisms.com/blog/the-history-of-design-model-making-and-cad

Historical milestones

https://engineering.rowan.edu/_docs/civilenvironmental/cee-materials-reading-assignment.pdf

Capitol Complex, Chandigarh

https://www.archdaily.com/806115/ad-classics-master-plan-for-chandigarh-le-corbusier

Parametric-sustainable Design

https://parametric-architecture.com/8-spectacular-parametric-designs-generating-sustainable-architecture/

Modular construction

https://99percentinvisible.org/article/modularity-modern-history-modular-mass-housing-schemes/

Leaning tower correction

https://leaningtowerpisa.com/facts/how-pisa-leaning-tower-was-stabilized

High-speed railways

https://www.youtube.com/watch?v=XAIwcJvKEvs

REFERENCES AND SUGGESTED READINGS

Books and Publications

- [Book] Fletcher, B. (1896). *A History of Architecture on the Comparative Method*. Batsford.
- [Book] Stearns, P. N. (2020). *The industrial revolution in world history*. Routledge.
- [Book] Antonelli, M., & Della Vecchia, G. (2022). *Civil and Environmental Engineering for the Sustainable Development Goals: Emerging Issues* (p. 108). Springer Nature

Web sources

- https://www.dezeen.com
- https://www.worldhistory.org/timeline/architecture/
- https://engineering.rowan.edu/_docs/civilenvironmental/cee-materials-reading-assignment.pdf

03 INFRASTRUCTURE

UNIT SPECIFICS

Through this unit we have discussed the following aspects:

- ***The present and future projection of various facets of Infrastructure development***
  - *Habitats–Megacities, Smart Cities, futuristic visions*
  - *Transportation–Roads, Railways & Metros, Tunnels (below ground, under water); Seaports, River ways and canals; Airfields and Airport, Futuristic systems*
  - *Energy generation–Hydro, Solar, Photovoltaic, Wind, Wave, Tidal, Geothermal,*
  - *Thermal energy and New sources*
  - *Water resource management*
  - *Telecommunication–Towers, above-ground and underground cabling*
- ***Awareness of various Codes & Standards governing Infrastructure development***
  - *ISO Codes for infrastructure construction*
- ***Innovations and methodologies for ensuring Sustainability in Infrastructure***

Besides giving a large number of multiple choice questions as well as questions of short and long answer types marked in two categories following lower and higher order of Bloom's taxonomy, a list of references and suggested readings are given in the unit so that one can go through them for practice.

There is a "Know More" section, which has been carefully designed so that the supplementary information provided in this part becomes beneficial for the users of the book. It is important to note that for getting more information on various topics of interest some QR codes have been provided which can be scanned for relevant supportive knowledge This section mainly highlights applications of the subject matter for our day-to-day real life or/and industrial applications on variety of aspects, case study related to environmental, sustainability, social and ethical issues whichever applicable, and finally inquisitiveness and curiosity topics of the unit.

RATIONALE

This fundamental unit discusses the important facets of infrastructure for the better understanding of its practical aspects during practice of Civil engineering, in the context of Sustainable development and future trends. It further emphasizes the role of standards and codes, and highlights the innovations and methodologies the civil engineer can leverage to develop sustainable infrastructure.

UNIT OUTCOMES

List of outcomes of this unit is as follows:

U3-O1: *Knowledge on the present and future projection of various facets of Infrastructure Development: Habitat, Transportation, Energy generation, Water provision and Telecommunication*

U3-O2: *Understanding/comprehension of various Codes & Standards governing Infrastructure development*

U3-O3: *Knowledge on innovations and methodologies for ensuring Sustainability in Infrastructure development*

Unit-3 Outcomes	EXPECTED MAPPING WITH COURSE OUTCOMES (1- Weak Correlation; 2- Medium correlation; 3- Strong Correlation)						
	CO-1	CO-2	CO-3	CO-4	CO-5	CO-6	CO-7
U3-O1	3	3	3	-	3	2	1
U3-O2	1	1	1	2	1	1	3
U3-O3	2	1	2	1	2	3	2

The term 'infrastructure' can be traced to 1927 referring to the various systems, amenities and facilities of public works that enables the provision of services. For example, Transportation services is provided by the associated infrastructure of roads, rail lines, harbours, bridges, etc. and Irrigations and Water management service is provided via dams, canals, etc. Today, classified as 'hard' infrastructure, these physical systems support various other services such as, waste management, telecommunication, power generation.

Prud'homme (2004) enlisted the **common characteristics of infrastructure**, as follows:

i. Infrastructure "*are capital goods*", i.e., it is a means to provide service and not consumed directly, requiring labour and other inputs for it to be useful.

ii. It is "lumpy" and not incremental, i.e., it is of a limited capacity to handle demand and cannot meet growing demands. It also requires years for being built and and of being in use.

iii. It is long lasting, having long term implications on maintenance and hence, financing.

iv. It is space specific and immobile, having further implications on financial capital.

v. Infrastructure and the service it renders is subject to market failures, decreasing costs, externalities, etc.

vi. It is used by both, households and enterprises, as it increases welfare and productivity.

In contrast, 'soft' infrastructure refers to the institutions that maintain health, socio-cultural and economic standards of a society, such as, healthcare and financial facilities, law enforcement and education. Of these, some are 'critical' infrastructure, identified as per the priority of the Nation. For India, these are Power, Telecom, Aviation, Energy, Banking, Cybersecurity and Disaster Management.

Another classification of infrastructure is based on its direct or indirect impact on production. The types that are essential for either improving or impeding production and distribution, such as, power, irrigation, transport, communication, etc., are classified as 'Economic Infrastructure, while those that aid economic progress, such as, Education, Healthcare and Housing, are classified as 'Social Infrastructure'.

3.1 HABITATS

Together, these various facets of infrastructure respond to the physical characteristics of a region to support the use of resources and productivity by the inhabitants beyond the boundaries of their natural habitat, with the underlying motivation for socio-economic development. In the following Sections, the present and future projection of various these various facets are discussed.

European Nature Information System (EUNIS) defines a **habitat** as "*a place where plants or animals live normally, characterized primarily by its physical features – topography, plant or animal physiognomy, soil characteristics, climate, water quality, etc., and secondarily by the species of plants and animals that live there*". In 2004, EUNIS classified various habitats, such as, marine, coastal, inland surface water, grasslands, woodlands and forests, along with recognising habitats created out of human intervention, such as, cultivated agricultural, horticultural and domestic habitats, and constructed, industrial and other artificial habitats, such as cities.

3.1.1 Megacities

In 2018 "*World Urbanisation Prospects*" the UN identified very large cities with population of over 10million as '**Megacities**' and UN further enlisted 33 megacities of which half of these urban agglomerations are in India and China and 27 are in the developing regions of the world, termed as the 'global south'. Globally, this is projected to rise to 43 by 2030, and the number of cities with 1 to 5 million inhabitants is projected to grow to 597. The characteristics of a Megacity are; *size, rate of growth and complexity* in terms of administration and infrastructure (Wenzel et al., 2007), and are the economic drivers of the country.

The urban population is projected to increase up to 28% worldwide, while rural population is projected to fall to 40% by 2030 (UN, 2018). Tokyo, Delhi and Shanghai that ranked 1st, 2nd , 3rd back in 2018 has already been overtaken by Shanghai, Chongqing, Beijing and Guangzhou, ranking Delhi at 5th today. This trend of fast-growing cities is widely noticeable in Africa as well, with Lagos in Nigeria being among the fastest growing, and other cities like Kinshasa in Democratic Republic of Congo and Dar Es Salaam in Tanzania.

Li, et al. (2019) analyses **four major problems of Megacities** from urban geography and ecology perspective, as follows;

- ***Land subsidence*** due to over exploitation of groundwater.
- ***Environmental problems*** such as, pollution, urban heat islands, urban air quality and haze, carbon emissions and dust storms, etc.
- ***Traffic congestions***, parking difficulty, public transport. This is closely related to the above problems of pollution, haze, etc.
- ***Energy consumption and production*** that is inefficient and unsustainable.

However, there is also the problem of declining of urban population. While large influx of population into urban boundaries is on rise, most of these cities also face high risk of population decline and loss of life due to geographical location, mostly coastlines, and consumption patterns are the vulnerable to at least one of the 6 types of natural disasters – cyclones, floods, droughts, earthquakes, landslides and volcanic eruptions. Some of the factors that contribute to the high risks of megacities in developing nations are; high population exposure due to housing concentration, complex and ageing infrastructure, lack of robustness of critical facilities and weakness of preparedness for response and relief (Wenzel et al., 2007). In addition, stagnating population has been associated with low fertility rates, particularly in Europe.

The mass movement of population from rural to urban has led to drastic changes in the constructed habitats and has put ever increasing demands and stress on infrastructure, limited resources and inequitable access, and deteriorating living conditions, leading to a large chunk of the population to be below poverty line and without social welfare. While cities today approximately occupy only 2% of the total land, it contributes towards 70% of the GDP and is responsible for 60% global energy consumption, 70% greenhouse gas emissions and 70% global waster generation. The enormous requirements of natural and human resources for energy, food, industrial production, services, construction, infrastructure, and maintenance, causes severe ecological impacts. In turn, development sprawl encroaches ecologically sensitive areas, such as, estuaries, coasts, riverbeds, which in turn require high developmental costs and maintenance more expensive. Other anthropogenic hazards, such as terrorism and war are also largely targeted towards megacities.

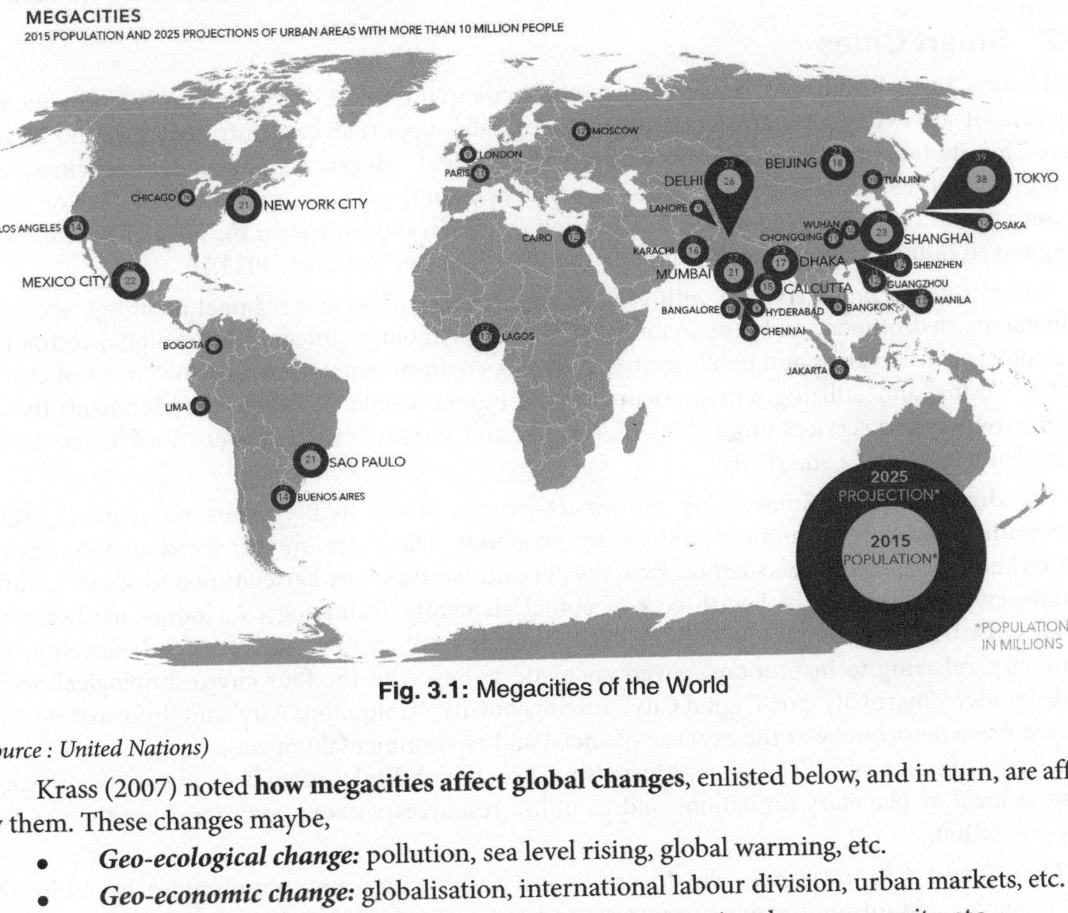

Fig. 3.1: Megacities of the World

(source : United Nations)

Krass (2007) noted **how megacities affect global changes**, enlisted below, and in turn, are affected by them. These changes maybe,

- ***Geo-ecological change:*** pollution, sea level rising, global warming, etc.
- ***Geo-economic change:*** globalisation, international labour division, urban markets, etc.
- ***Geo-social change:*** urban disease, social justice, migration, human security, etc.
- ***Geo-cultural change:*** urban ethnicity, social movements, global media, urban cultural diversity and hybridity; and
- ***Geo-political change:*** resource security, social stability, social justice, welfare, etc.

To combat these challenges, the '*New Urban Agenda*' was adopted during the UN Conference on Housing and Sustainable Urban Development (Habitat III) in 2016, that outlined the implementation of; Urban rules and regulation, Urban planning and design, and Municipal finance, with consideration of national urban policies to support sustainable development. In addition, megacities are greatly leaning on digital Information and Communication Technologies (ICT), to envision the ever-growing city through 3D modelling and visualisations, image and data fusion, Big Data, IoT and real-time Earth observations, becoming the backbone for 'digital cities of the future' or '**Smart cities**'. However, while megacities are heavily into employing 'Intelligent' strategies, but it may be argued that without addressing the elements of sustainability and noted improvement of quality of life, and only through employment of ICTs is inadequate. Presently, western megacities such as London, Paris and New York, are focusing on infrastructure for quality of life by ensuring; sustainable utilisation of land resources, improvement of design quality of buildings and urban spaces, balance between economic growth and environmental protection, improve air and water quality, climate change, etc. This gives an insight into what should the priority of developing nations such as India be to achieve SDG 11 – Sustainable Cities and Communities.

3.1.2 Smart Cities

SDG 11 aims for inclusive, safe, resilient, sustainable cities and 'smart city' planning is a proponent of the same, as it strives develop frameworks to technologically support all basic infrastructure and services required for its inhabitants towards becoming self-sufficient. Megacities and upcoming cities, with growing issues of sustainability are struggling to preserve natural and economic resources, lean on 'smart city' concepts however, the ***ground implementation is heavily dependent on the level of development, willingness to change and availability of resources*** (Bordoloi and Acharya, 2023).

'Smart' city is a broad concept with various sub-themes; urban and regional planning, economic development, environment and sustainability, ICT and technology. Integration of digital technology with improving urban areas and public spaces, reducing environmental impact, involvement of citizens in policymaking, and utilizing entrepreneurship and human capital for urban development; thereby making networks and services more efficient, flexible, and sustainable for the benefit of its residents is the key characteristic of a Smart city.

The **dimensions of Smart City** are smart–people, smart living, smart government, smart transportation, smart environment and smart economy. These are further expanded to include smart technology, smart infrastructure, smart water and waste, smart agriculture and smart security. In summary, the term 'smart' has three conceptual elements: Technology, including hardware and software infrastructures; People and their associated attributes of creativity, diversity and education; and Community, referring to institutions, governance and policy. And the four city technological brands included under 'Smart City' are 'Digital City', 'Intelligent City', 'Ubiquitous City' and 'Information City'. However, these must not be at the expense of social and environmental impact as 'quality of life' is the eventual goal of smart cities. Therefore, a term **'smart sustainability'** is often used to stress on the need to respect local or planetary limitations and to utilise resources without compromising the needs of future generation.

Thus, Smart City strategies help to create sustainable cities and communities by addressing social problems, optimising financial resources, and mitigating environmental consequences through conservation and controlled use of natural resources. **Key features of 'sustainable smart cities'** include *compactness, population density, sustainable transport, mixed land use, green areas, passive solar design and diversity*. However, there are still gaps between real world problems and available strategies and solutions.

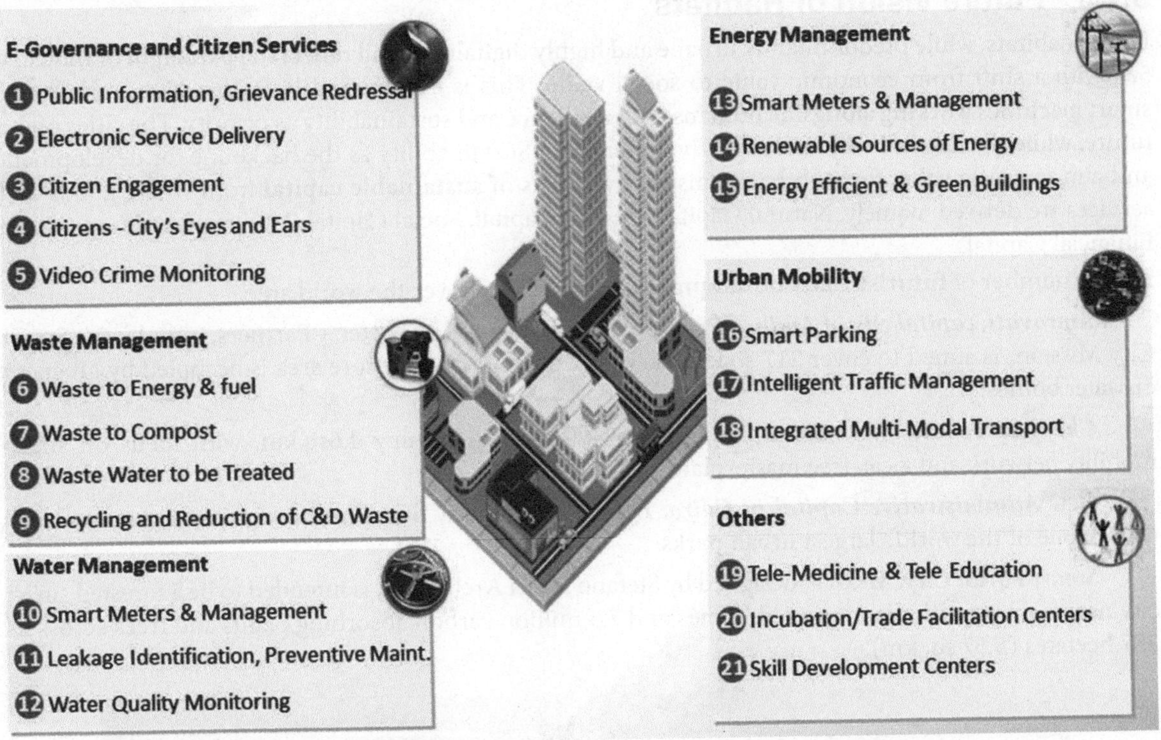

Fig. 3.2: Illustrative List of Smart Solution, Mission Statement & Guidelines, MoUD, GoI (Jun 2015)

In June 2015, the Govt. of India launched the '***Smart Cities Mission***' to develop sustainable and inclusive cities, keeping in mind that by 2030, 40% of the Indian population will be in urban areas and will contribute towards 75% of the GDP. The Ministry of Housing and Urban Affairs (MoHUA) outlined the strategic components of a smart cities as; adequate water supply, assured electricity supply, sanitation and solid waste management, efficient public transportation and urban mobility, affordable housing, robust IT and digitalisation, e- governance and citizen participation, sustainable environment, safety and security of citizens, and health and education. The mission envisioned 100 cities across the country and strategized based on area-based development through retrofitting, redevelopment, greenfield development, and pan-city initiatives. There are several schemes, such as, Atal Mission for Rejuvenation and Urban Transformation (AMRUT), Swachch Bharat Mission, National Heritage city Development and Augmentation Yojana (HRIDAY), Digital India, Skill Development, Housing for All, etc., that offer complementarity to the Mission and convergence of these is promoted. Three tiers of monitoring, at national, state and city level is proposed, and while it is presently centrally funded, it requires state government and urban local bodies to contribute equal amounts for implementing the Smart City, approximately 100crore per city per year for 5 years. The total allocated investment stood at Rs.205,018 crore ($ 27.6 Billion) as of March 2021 and a recent Rs.16,000 crore has been allocated in Budget 2023. Presently, MoHUA reported that more than two-thirds of the total 7,804 projects have been completed.

3.1.3 Future vision of Habitats

Future habitats, while predominantly urbane and highly digitalised, will reflect the paradigm of Industry 5.0, with a shift from economic value to social value. This is exemplified by automation, robots and smart machines working alongside humans with resilience and sustainability as priority. The cities of the future, while 'intelligent', will lean strongly on tenets of Sustainability as the backbone for development and aim to create value through harmonising **five types of sustainable capital** from where goods and services are derived, namely, Natural Capital, Human Capital, Social Capital, Manufactured Capital and Financial Capital.

A large number of **futuristic city developments** underway all over the world are;

Amravati, capital city of Andhra Pradesh, India, designed by Foster + Partners, is under the Smart City Mission, is aimed to cover 217 sq.km. such that over 60% of the core area is occupied by greenery or water bodies.

Chengdu Future City, China, designed by OMA, will occupy 4.6sq.km. with focus on smart mobility network and a car-free masterplan.

New Administrative Capital in Cairo, Egypt, designed by SOM, will cover 700 sq.km. and will feature one of the world's largest urban parks.

Smart Forest City, Mexico, designed by Stefano Boeri Architectti, is intended to be a forested smart city near Cancun with plant covered homes and 7.5 million carbon absorbing plants and trees across its 557 hectares (5.57 sq. km).

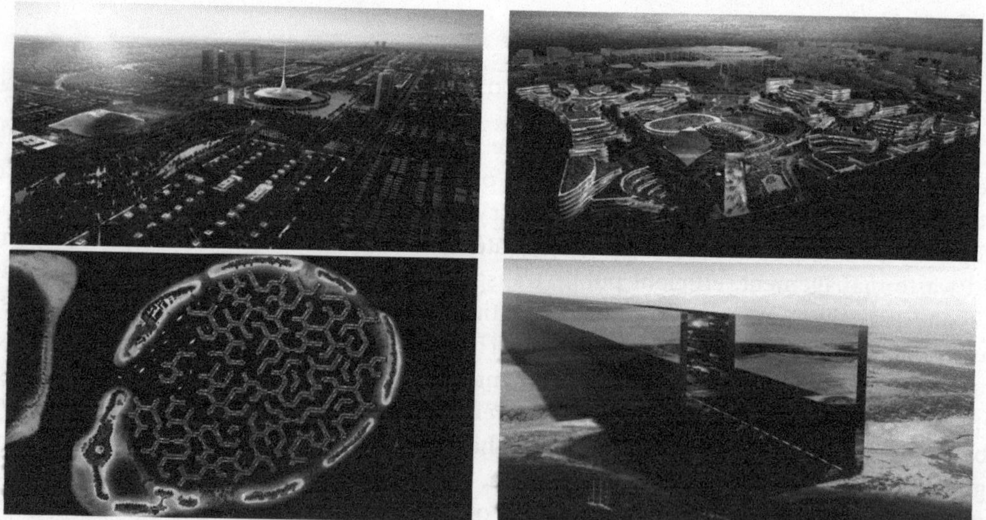

Fig. 3.3: Future Projects–(*Top Left*) *Amaravati*, India, by Foster+Partners, (*Top Right*) *Chengdu Future City*, China, by OMA, (Bottom Left) *Maldives Floating City*, Maldives, by Waterstudio, (*Bottom Right*) *The Line*, KSA, by Neom

Several attempts to use advanced building and construction technology is also underway to **reclaim or rebuild presently vulnerable cities**, such as;

BiodiverCity, Malaysia, also designed by BIG, is a 1821 hectare (18.21 sq.km) development of three artificial islands built of the shore of Penang Islands, built using combination of bamboo, timber and concrete.

Maldives Floating City, Maldives, designed by Waterstudio, as a response to the rising sea level and the concern of the Maldives becoming uninhabitable by 2050, will be built on hexagonal structures that rise with the sea.

A prototype climate-resilient floating city is also being designed by BIG and tech company, ***Oceanix, off the sea of Busan, South Korea.***

Further, state of the art, **breakthrough proposals** are also underway in unoccupied land stretches with the intention to design for the future population and a sustainable way of living, such as;

The Line, Saudi Arabia, which will house a 170km long, 500m tall, but 200m wide, linear city stretching to touch the sea, and is envisioned as a model for 'nature preservation and enhanced liveability'.

Telosa city, USA, designed by Bjarke Ingels and BIG on an unoccupied 150,000-acre (607.03 sq.km) site in the western desert.

Design of green buildings, use of eco building materials, incorporation of natural environment and urban open, green spaces, apart from the 'smart' services and their allied infrastructure would be characteristics of cities of the future. In addition, there is a hope that the citizens of the future will be more conscientious as a society and continue to strive towards the SDGs collectively. This brings us back full circle to the early '**social indicator movement**' of the 1960s where the important linkages between infrastructure and quality of life (Terleckyj, 1975) was captured in order to systematically assess existing possibilities for social change, by mapping the infrastructure against attributes of human habitat, i.e., Health, Safety, Recreation and Aesthetics, Economic opportunity and Leisure. Aschauer (1989) reflects that, *the general public desires cleaner environment, safer urban streets, increased mobility and economic opportunity for the disadvantaged, and an economy well equipped to compete internationally*; and empirically establishes the importance of infrastructure development and investments towards improved quality of life and economic performance.

3.2 TRANSPORTATION

The global economy functions through trade and commerce, and transportation service sector is the most essential aid. It not only supports the supply chain through timely movement of raw materials and resources, but also enables eventual distribution of finished goods, in turn assuring economic competitiveness and growth. It is also the mode of travel for people for employment and education opportunities, healthcare, leisure, or social networking, etc. But while this was greatly hampered in the pandemic era, with millions with home bound. The times are also exemplar of the heights of exploitation of transportation infrastructure and logistics, becoming one of the most critical infrastructure sectors along with Energy, Water and Telecommunication. And today's new normal is quick and easy, door-to-door delivery, demanding development of further land, water, underground, air, as well as futuristic modes of transport infrastructure.

3.2.1 Land Transport Infrastructure

Land transport can be broadly categorised into roads, railways and metros; a 'way' for travel. The word 'way' as per Britannica, stems from the Latin '*veho*' which means, "*I carry*", derived from the *Sanskrit* word, '***vah***' meaning to "*carry, go or move*".

Roads

The generic word 'road' used to encompass all land vehicular ways is derived from the Old English word '*rad*', which means "*to ride*". However, the term highway predates it, back to the elevated Romans roads which created a mound in the centre and ditches on the side, upon compacted layers of soil and gravel. It is used to refer to major roadways that connect several rural and urban space and is characterised by various controlled points of entry and exits for traffic. While 'streets' has its origin in Latin '*strata*' meaning paved, and is till date used to refer to all small rural scale roads.

Road networks follow a system of hierarchy in their capacity and capabilities. Most roads world over are conventional, undivided two-way; however, there are divided roads such as; expressways, having minor at-grade intersections; freeways, having no at-grade intersections, collectively called motorways in the UK. Functionally roads are classified as: Highways–the major roads between regions; Arterial roads – that carry through traffic from adjacent areas and are major roads within a region; Collector, distributor and feeder roads – that carry only through traffic from a certain area perimeter; and local streets – that do not carry through traffic but only serve adjacent properties.

The earliest records of animal trodden paths used by man have been dated back to 6000 BCE, with evidence on first constructed road dating 4000BCE at Ur, present day Iran. Oldest existing paved road, made of layers of sandstone bound by clay-gypsum mortar with two rows of basalt slabs in the centre for use by foot, while the dipping edges were for animals, was built by the Minoans on the island of Crete. It was about 50km long at an elevation of 1300metres, from Gortyna to Knossos. The earliest long-distance road, traversing 2500 kms, between Susa in the Persian Gulf to the Mediterranean ports of Smyrna (Izmir) and Ephesus, was first used around 3500 BCE but came into organised use aroud 1200 BCE under the Assyrians. The Egyptians around 2600 BCE built roads to haul limestone blocks for pyramids and to connect towns and temples. Routes from Thebes and Coptos on central Nile ran towards the Red Sea and from Memphis (Cairo) towards Asia Minor. By 1500 BCE, a trading network across eastern and central Europe called **the 'amber routes'** grew for trading various materials such as, amber from northern Europe, salt from Austria, freestone from Belgium, lead and tin from England, flint from Denmark,etc. and connected Hamburg, Cologne, Frankfurt to Passau and Venice. In 334 BCE, Romans championed road making building nearly 85,295 kms, radially connecting Rome to its several provinces, with 29 great military roads (*viae militares*). **The Appian Way** was the most famous one, constructed in 312 BCE and at a distance of almost 660km from the capital, along the Mediterranean coast. In addition, the Romans also had two classes of public transport – the express service and freight service. China too had an extensive road system by the 7th century CE across 40,000km, development of which began under the Emperor Shinhuangdiin 220BCE. In India, under the Mauryan Empire around the 4th century BCE, had quality roads across its empire which stretched from Indus River in the Northwest to the Bramhaputra in the East, from Himalayas in the North to the Vindhyas in the South. Their **Great Royal Road** ran through Taxila or Takshashila, in modern day Afghanistan, across Punjab and Prayagrai, until the mouth of the Ganga River. The famed **'Silk route', the longest road of the time** that linked the Mediterranean with China, is said to have existed partially around 300 BCE and by 100 BCE it was pivotal to trade. However, the following centuries saw neglect of road systems without strong dynasties and leaderships, and wasn't until 8th century that road revival became a priority. In 9th century that the Moors of Cordoba, Spain developed an extensive street network. The following 11th and 12th century witnessed revival of cities and roads, mostly in western Europe, and by 15th century, well maintained infrastructure was ubiquitous. Meanwhile, in South America, the Inca empire (1000 – 1450 CE) extended the '*Qhapaq Nan*' or royal road from Ecuador, Peru until Chile, with two parallel systems – one along the coast and another along the Andes. Eventually, by 18th century, the modern masters of road building – Treaguest, Telford, Mc Adam, Mitchell, appeared in England and France.

Road design is an exceedingly important aspect of national and regional planning, especially in the context of urban connectivity and the population, commerce, industry and transportation needs of the community. Estimating traffic on a route as well as conducting civil surveys to establish the site conditions are integral to successful road system planning and design. To design a road, three-dimensional road alignment of the cross-sectional profiles along the terrain on which it is laid is of utmost importance. For quality and uniformity, **standards and codes** are established for different types of roads. It helps with guidelines on, number of traffic lanes as per traffic volume and speed, width of lanes, carriageways and shoulders, and specifications on roadside barriers. Other integral elements of road system design include design of pavements and drainage. Safety and serviceability of the roads are also critical aspects to be considered and financing and maintenance are key determinants. Jurisdiction of the section of the road and its onus, also plays an important role in case of any legality and taking responsibility. In India, roads are under the aegis of Ministry of Road Transport and Highways. Some of the **salient features reported** in the review 2022, note the major push in the following areas;

- Highway development
- Connectivity
- Logistics development
- Online citizens service
- Under Bharat series, regularisation of registration mark
- Retro fitment of CNG and LPG kits allowed in BS VI
- Bharat N-CAP safety rating introduced
- Compensation of victims increased
- Green fuel and vehicles
- Construction of '*amrit sarovar*' along National Highway
- NHAI InvIT bonds enlisted on BSE and NSE, and
- Manthan conference held in Bengaluru to discuss issues in road, transport and logistics.

Railways and Metros

In earlier chapters, we have discussed the oncoming of railways. The role civil engineer is pivotal for railway transport infrastructure, and encompasses surveying for a new line, construction and maintenance of the line, ensuring longevity, safety and reliability of the structure. In addition, the civil engineer also has to design bridges over rivers, station buildings and allied facilities, such as, office rooms, parcel offices, goods sheds, restrooms, wating lounges, as well as, loco sheds, pump houses, water and drainage lines, etc.

Railways evolved from **'tramways'** which were originally of stone slabs, timber and baulks, laid in flush with the road surface for horse carriages, and later, reinforced with iron straps or plates. Further on, these were improved, and tracks were designed, having angle irons with a vertical leg, later replaced by cast iron beams, which further evolved into rail sections on which the locomotive's wheels align. The wheels transfer the load of the locomotive on to the two rails of the track, kept at specific distance, i.e., gauge, and is placed on perpendicular sleepers equitably distanced on a bed of ballast. There are ballast-less, and continuous longitudinally- supported tracks as well, the latter being very uneconomical. Earlier, rails were of 'I' or dumb- bell sections; but this further developed into a 'T'-section, laid inverted and popularly termed as the 'Flat-footed' rails. Other rail profiles are Bullhead rail, Grooved rail, Bridge rail (inverted U-shape) and Barlow (inverted V-shape) rails. There are two common types of gauge– Broad gauge (1676 mm) that supports the speed of 100-160 Km/h, and Metre gauge (1000 mm) having a maximum permitted speed of 75Km/h. Another type is Narrow gauge, such as those of the Darjeeling Himalayan Railway or the Toy train which has achieved the UNESCO World Heritage status. Sleepers maybe made of wood, cast iron, steel, concrete. The rails are fastened on to the sleepers with dog spikes, and the ends are connected by fish plates or fish bolts, and can be switched to direct the trains. Other types of fastening are–round spikes and screw spikes, different types of bearing plates, tie bars, etc. The rail line is either placed on an embankment, or in a cutting where a pit is dug and the centre is raised to house the track, as it cannot be laid directly on ground. The railway track comprises of rails, sleepers, fastenings and ballast, and may also be called **'permanent way'**.

The **first steam locomotive** to carry passengers, designed by engineer, George Stephenson, began operations between Stockton and Darlington in 1825. The first railway proposal in India under the British rule was made in 1832, and the **first train transport named 'Red hill'** plied in 1987 carrying freight of granite for road -building. India's inaugural passenger train operated by the Great Indian Peninsular Railway, having three steam locomotives named *'Sahib, Sindh and Sultan'*, ran between Bori Bunder (Bombay) and Thane, for a distance of 34kms, in 1853, and first passenger train ran between Howrah and Hoogly in West Bengal in 1854. Today, the Indian Railways is Asia's largest network and among the world's largest. It is about 108,706 track Kms and runs around 11000 trains daily, off which 7000 are passenger trains carrying around 13 million passengers every day.

In 1984, South Asia's first subway line began operations in Kolkata after 23 years since the commencement of underground construction. Delhi, too, had a urban rapid transit proposal back in 1969 but the construction began in October 1998. Today, the Delhi Metro is a benchmark, not only as the largest and busiest metro system, but also due to its state-of-the-art design and construction. The network consists of 10 colour coded lines, covering the National Capital region and its satellite cities of Ghaziabad, Faridabad, Noida and Gurgaon. It is a mix of broad and standard-gauge and has underground, at-grade (at road level) and elevated sections. **DMRC (Delhi Metro Rail Corporation)**, a company with equity from Government of India and Government of Delhi, under leadership of E.Sreedharan, has been certified by U.N. as the first rail-based system in the world to get carbon credits for reducing greenhouse gas emissions and carbon emissions. Mumbai also boasts of the first of its kind, Mumbai Monorail, that wascompleted in 2014 and is presently witnessing the development of the sixth longest operational metro network in India.

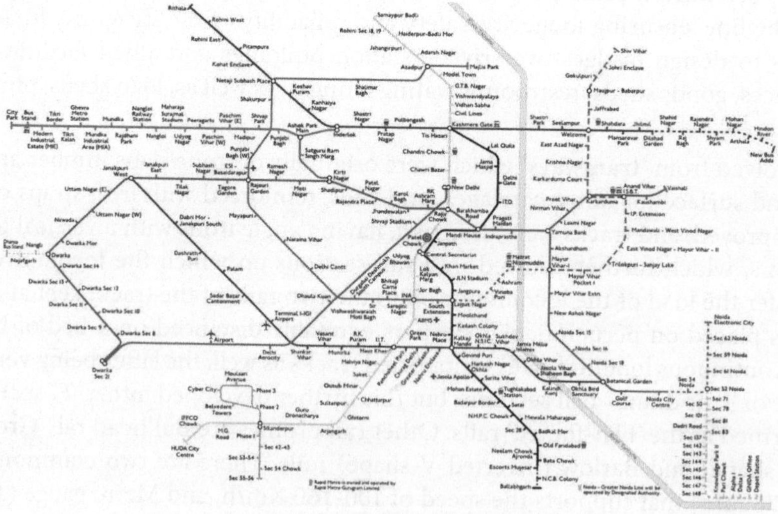

Fig. 3.4: Delhi Metro Map

(source : India Tourism, 2022-23)

Tunnels : Below ground or under water

Tunnels are under ground or under water, enclosed passages constructed by digging through rock, soil and earth, for movement of vehicles. With the massive excavation needed, the integrity of the passage is subject to the loading and pressure of the surrounding earth and/or water, construction of tunnels is highly specialised. There are various methods of tunnel construction on land, such as, *Cut and cover method* where a trench is cut and covered by some support;, *Bored method* where boring machines are used; *Clay kicking* method common from the time of World War I, where clayey soil is literally kicked to create a tunnel; *Shaft method* is used for reaching great depths from the ground level and pre-cast shaft structures with concrete walls are lined and connected; *Pipe jacking method* where pipes are driven underground using hydraulic jacks, usually under existing infrastructures like roads, railways, etc.; and *Box jacking method* where boxes instead of pipes are driven in. For underwater tunnel construction usually calls for immersing tubes or boring tunnels through rock.

The **world's longest road tunnel** is the *Lærdalstunnelen*, of 24.5 Km linking Aurland and Lærdal, and offering a ferry free connection built in 2000, followed by the Yamate tunnel in Tokyo, Japan measuring 18.2Km built in 2015 and the Zhongnanshan in Shaanxi, China measuring 18.04Km built in 2017. The Eisenhower Road tunnel in Colarado is one of the **world's highest tunnels** at a height of 3,401m above sea level, on the Rocky Mountains.

The **world's longest and deepest train tunnel** is the Gotthard Base Tunnel in Switzerland, running 57 Km in length and at a depth of 2300m under the Swiss Alps, significantly reduced train travel time connecting Zurich to Milan. Previously this distinction was held by the Seikan railway tunnel, spanning the Tsugaru strait connecting Honshu and Hokkaido, running 53 Kms at 140m depth. Another phenomenal feat is the Channel tunnel across the English Channel between England and France, spanning 50Km, having two rail and one service tunnel for passenger and freight.

A piece of modern engineering is the **SMART** (Stormwater Management and Road Tunnel) Tunnel in Kuala Lumpur, Malaysia, designed keeping in mind the flash flood situation that the city faces and operates in three ways : as a road tunnel when there is no flood, the upper level operates for traffic while the lower channel allows water diversion when there are medium floods; and closes for traffic but allows floodwater to flow through via a holding pond, bypass tunnels and a storage reservoir during heavy floods.

3.2.2 Water Transport Infrastructure

Seaports

Ancient Greeks heavily relied on sea travel as land travel was difficult. Greece is a series of archipelagos and peninsulas, surrounded by the Ionian, Mediterranean, and Aegean Sea, which they traversed by trireme ships. Beyond travel, they instituted maritime trade and frequented the ports of Canopus before Alexandria in Egypt and Messina in Sicily from Athens. Ostia Antica and Portus were later set up in Rome, and Swahili kingdoms of East Africa were described to have vibrant trade ports. Trade flourished amidst the Arabs, Greeks, Romans, Egyptians, Africans and Chinese. The south-western coastal ports of Muziris and Calicut, or Kozhikode, in present day Kerala, are accounted in ancient texts and is to have played a crucial role in the spice trade. The latter gained prominence after the arrival of Vasco da Gama. The port of ***Lothal***, the southernmost city of the Indus civilisation, at the edge of the Arabian Sea in present day Gujrat and is believed to be **the world's oldest dock**, dating back to 2200 BCE. Another important port in Gujrat, at the mouth of the river Narmada is Bharuch. East of the Indian subcontinent, Chittagong in present day Bangladesh, has been found referred in Ptolemy's map, dating back to 2nd century. And the south-eastern ports of Tuticorin (Thoothukudi), Arikemedu at modern day Pondicherry and Poompuhar, all in Tamil Nadu were bustling ports. In the far east, ancient seaports were Guangzhou in China during the Qin Dynasty and Osaka in Japan during the Edo period.

Civil engineering plays a crucial role in the design of water transportation and construction of ports and harbours. Ports are locations where ships and vessels can dock and allow movement of people and goods and are either on a coasts or shores, while harbours are constructed for the safe keeping of the vessels. Other infrastructures that have nuanced variations but support sea travel are; quay or wharf, pier, jetty , berth and dock, and all together form a network of waterway infrastructure for overseas transportation.

Inland waterways and canals

The beginning of rails marked the demise of canals and inland waterways, which used to be the preferred mode of transportation. All ancient civilisations settled around rivers, such as, the Indus civilisation along Indus River, Mesopotamia between the Tigris-Euphrates, Memphis (Egypt) along the Nile. Several prominent European capital cities, such as London, Paris and Amsterdam all are also along rivers. Apart from being a source of water for consumption and irrigation, riverways and inland waterways became the mode of trade and travel. Several other cities, such as, Giethoorn in Netherlands, Birmingham in England, Burges in Belgium, Hamburg in Germany, Stockholm in Sweden, are all traversed with network of canals with most prominent being **Venice**, Italy, which has **150 canals** including the *Grand Canal*. These were built by lining the dugout pits by closely spaced alder wood which is waterproof, to make the lagoon fit for habitation. While most of these canals are 1.5-2m deep, the *Canale della Giudecca* that separates the main part of Venice from the island of Giudecca, is about 12-17m deep.

Two of the most ambitious canal projects that are engineering feats are; the **Suez canal**, also referred to as *Qanat al Suway* (length of 193Km with branches, depth of 20m and width of 205m) across the Isthumus of Suez in Egypt which connects the Guld of Suez and the Mediterranean, thereby allowing a quicker path between the Pacific and Indian Oceans, and the **Panama canal** (length of 82Km with branches, depth of 12m and width of 150m) which connects the Pacific and the Atlantic Oceans across the Isthumus of Panama amidst the Caribbean. These two interventions significantly changed the time required and ease of transport for trade, becoming pivotal contributors to economic growth.

The **world's oldest and longest man-made canal** is the '*Great Canal*' or the Beijing- Hangzhou canal that connects the Yangtze and Yellow He rivers, across 1,782 – 2,470 Km with branches in length and varying width between 40-350m, having a depth of 2-3m. It is adorned with 21 gates and 60 bridges and has been recognised by UNESCO In 2014.

3.2.3 Air Transport Infrastructure

Aeroplanes, helicopters, light aircrafts, hot air balloons, blimps and gliders, drones or UASs (unmanned aircraft systems) are all vehicles of air transport and require certain physical infrastructure to support flight and while not in flight, such as service, maintenance and parking.

Airports are complex transportation hubs and is spatially divided into three: airside, landside and terminal that connects the two. The civil engineer has the responsibility of structurally designing and constructing; the airside layout comprising of runways, taxiways, parking aprons, lighting and signages, navigational and visual aids; the landside facilities, such as parking lots, fuel tank farms, access roads, technical buildings like control towers for ground aid, etc. and passenger and cargo terminals.

Helicopters too have designated bases with fixed operations and services like customs, fuel bunkering and maintenance, called heliports. It is like a small airport specifically for helicopters and other vertical-lift vehicles. Often high-rise buildings, hospitals and other buildings or campus of importance have helipads for landing and take-off only. This infrastructure is of particular importance as helicopters and drones are the preferred mode of transport during natural disasters for rescues and searches, supply as well as surveillance.

Aviation falls under critical infrastructure in India and has been a hot target for attacks, online and offline. The regulatory body for India is the Directorate General Civil Aviation (DGCA), empowered by the **Aircraft Act 1934**, implements standards and recommended practices of the International Civil Aviation Organisation (ICAO), and later further bolstered by the Aircraft Rules 1937, is authorised to specify requirements and compliance procedures through Civil Air Regulations (CAR). It outlines the operations and planning of infrastructure as well. The DGCA also regulates the airspace and in turn, monitors and supports the use and manufacture of UAS, as well as authorises remote pilot training and certifications, to overall ensure National security.

3.2.4 Futuristic Transport Infrastructure systems

'Smart' transportation is the vision of the future, with driverless cars, flying taxis, delivery drones and levitating trains being already on the horizon. **Novel concepts** of futuristic transportation, impose the need for reimagining the infrastructure, such as;

Self-drive Cars or autonomous vehicles (AV), a technology that is seemingly possible in the near future, require physical and technological framework to support and enable the operation of autonomous vehicles. The most important element is the need for communication network with real-time data exchange through vehicle-to-infrastructure (V2I) systems, such as sensors in roads or street signs that send signals to AVs, helping them navigate city streets. There are also alternatives like, dedicated short-range communications (DSRC) and cellular vehicle-to- everything (C-V2X) systems. Physical infrastructure is also crucial, such as, parking facilities equipped with autonomous parking capabilities, maintenance stations specifically designed for self-driving cars, and a comprehensive network of charging stations as many self-driving cars operate on electric power, to support uninterrupted operation.

Maglev, a technology where magnetic levitation actuated by two sets of opposing electromagnets along the tracks, allows trains to travel at high speeds. Shanghai Transrapid is the presently the world's fastest commercial electric train and goes up to 431 Km/h. Japan began development and construction of SCMaglev under the Central Japanese Railway Company in 1969, and the HSST (Linimo) in 1974. A new maglev line, named the Chuo Shinkansen started in 2014. In 2016, South Korea inaugurated the Incheon Airport Maglev. Currently, all maglev trains are in use in Asia.

Hyperloop, "*an ultra-high-speed transportation system in which passengers travel in autonomous electric pods*" futuristic transportation concept. It was first conceptualised by Elon Musk in a white paper in 2013 for intra-city mass transit and envisioned "*a tube, over or under the ground, that contains a special environment*". It consists of low-pressure tube with capsules that are transported at both low and high speeds throughout the length of the tube, such that the capsules are supported on a cushion of air and are accelerated by a magnetic linear accelerator affixed at various stations.

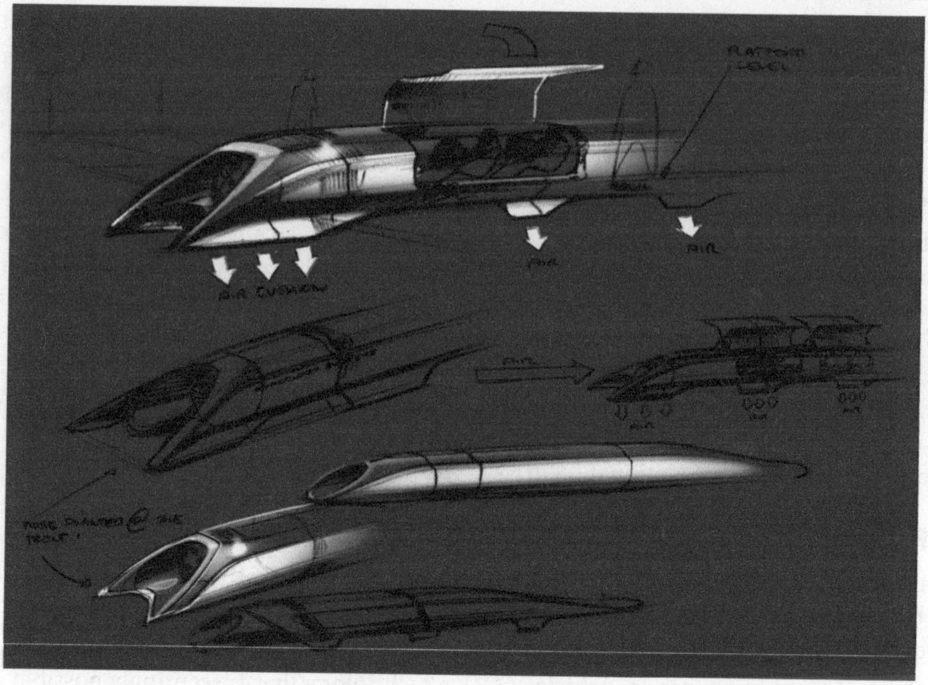

Fig. 3.5: Hyperloop Alpha concept sketch (source: www.tesla.com)

Underground tunnel roads are another vision of Musk for futuristic transportation where he conceptualises a network of underground tunnels equipped with trolley like platforms to lower the cars from the surface and speedily transport them to their destinations, alleviating congestion. This is much in line with the idea of autonomous vehicles, driverless cars, and flying taxis, which are all well underway in prototyping phase.

The motivations of these proposed futuristics concepts are; reduction of congestion and greenhouse gas emission, reduce fatalities and prioritise safety, provide access to disadvantaged and to improve on time. 'Quality of life' and low environmental impact are key and in turn, requires to re-imagine the supporting infrastructure for energy, telecom, etc.

3.3 ENERGY

Global energy infrastructure includes delivery systems of oil and natural gas pipelines, power generation facilities and transmission lines, storage facilities, etc. Inclusive economic growth, clean air, climate has necessitated innovations in efficiency gains, new production flows, digitalization, smart grids, etc., requiring infrastructure to become sustainable, resilient, and secure (International Energy Forum). Geopolitical and natural events and accidents may lead to disruption of energy markets, and hence, policy cohesion and regional market integration through cross-border networks are essential to meet SDG, i.e., Ensuring access to clean and affordable energy.

"*Energy generation depends on a country's natural resource endowments and technology to harness them*" as per Government of India Budget report 2012-13. India is the world's third largest producer, as well as, consumer of electricity, but this is predicted to increase by 4.5% annually to 2035. Presently, India's energy consumption is 1010kWh per capita against a world average of 3200kWh. Gujrat, Maharashtra,

Tamil Nadu, Rajasthan, Karnataka, and Uttar Pradesh are the top power producing states of the country. However, the key issues are poor transmission and distribution grids. India has both non-renewable reserves, of coal, lignite, petroleum, and natural gas, as well as renewable energy resources of hydro, wind solar, biomass, and cogeneration bagasse. But while the non-renewable resources are poor, India is richly endowed with renewable resources, and several policies and initiatives are presently in place to leverage that.

3.3.1 Renewable Sources

Renewable energy, popularly referred to as **'clean' energy**, comes from natural sources or processes that have a higher rate of replenishment than its consumption. Their sources are sunlight, wind, water, geothermal and biomass. While large hydroelectricity projects and biomass levy a trade- off on wildlife, biodiversity and climate change, other sources have very little negative impact on the environment and in turn, generate lower emissions.

Solar technologies can also provide heating and cooling solutions, natural lighting, and fuel for cooking. Solar power can be harnessed through mirrors and photovoltaic (PV) panels, that concentrate the solar radiation and converts it into usable fuel or electricity. PV cells are made of silicon and present technology has not only significantly reduced its prices, but extended its life to up to 30 years. Solar farms are fast becoming a large initiative where PV cells are splayed, often aided by mirrors to concentrate the sun light, and in some cases, wastelands, water bodies and wastewater facilities are employed.

Wind energy exploits the kinetic energy of moving air by using large wind turbines situated at on-shore or off-shore locations, as per the average wind speed of the location.

Geothermal energy is garnered through the heat extracted from the earth's interior reservoirs using wells and ducts. Hot fluid of varying temperature is pumped through turbines to generate electricity, and recycling of the used water and steam ensures low emissions.

Hydropower harnesses the energy of falling water from higher to lower elevations and can be generated from reservoirs or flowing rivers. Beyond providing electricity, these sources also provide water for irrigation and drinking, as well as help in flood and drought control, and navigation services. However, rate of precipitation and annual rainfall impacts the functioning of hydropower.

Tidal and wave energy of the ocean is also a potential source of clean energy and is generated through tidal barrages which are dam-like structures located in ocean bays and lagoons, or via devices which are ocean floor-anchored or placed just below the wave surfaces. These interventions may cause adverse effects on the life under water.

Bioenergy is produced from biomass which is the organic remains of various sources, such as, wood, charcoal, dung, manures, reside from agriculture and forestry, and other organic wastes. It is used mostly in rural areas for cooking, lighting, heating. However, in spite of its natural origin, it has detrimental environmental impact as it produces emissions and bioenergy plantations may lead to deforestation and land-use change, making it arguably not a 'clean' energy source.

India is the **third largest producer of renewable energy** and is rich in clean energy sources, such as, solar, wind, and small hydro, with high potential for energy generation. India globally ranked 4th in solar power capacity, wind power capacity and overall renewable power installed capacity as of 2021 and ranked 6th for hydropower generation in 2019. Today, as of January 2023, India's installed renewable energy capacity, is at 40.9% of the total installed power capacity. Solar (63.3 GW), hydropower (46.85 GW) and wind (41.9 GW) are the largest contributors followed by biomass (10.2 GW), small hydro (4.92 GW) and 'waste to energy' (0.52 GW) as contributors towards the renewable energy capacity.

The past year has been pivotal for development of renewable energy. The union cabinet approved 2,614 Crore investment in the 382 MW Sunni Dam Hydroproject by SJVN (Satluj Jal Vidyut Nigam). SVJN also signed an agreement with the Utter Pradesh government to implement 3 solar power project and commissioned a Solar power project (75MW) at Parasan Solar Park near Kanpur, U.P. They further signed up with Tata Power Solar systems to build a 1000MW solar project near Bikaner and announced collaboration with Govt. of Assam for development of hydro and renewable energy projects. National Hydroelectric Power Corporation (NHPC) and Govt. of Himachal Pradesh intend to implement a 500 MW hydroelectricity project in Chamba District, H.P., as well as two project in Nepal. Tata Power Green Energy Limited (TPGEL) commissioned a hybrid power project in Rajasthan and was awarded a solar project in Solapur, Maharashtra. National Thermal Power Corporation (NTPC) announced partial power generation from the floating solar energy plant in Kayamkulam, Kerala, and Jetsar, Rajasthan. Adani Solar and Smart Power India (SPI) signed an MoU to promote usage of solar rooftop panels in rural India. To further promote this the Ministry of New and Renewable Energy has developed a national portal for citizens to directly apply. Several **policies and programs**, such as, '*National programme on High Efficiency Solar PV Modules*', 'smart metre deployment' for National Smart Grid Mission, several electrification schemes and issues of sovereign green bonds and conferring infrastructure status to energy storage systems, are underway to improve India's power sector.

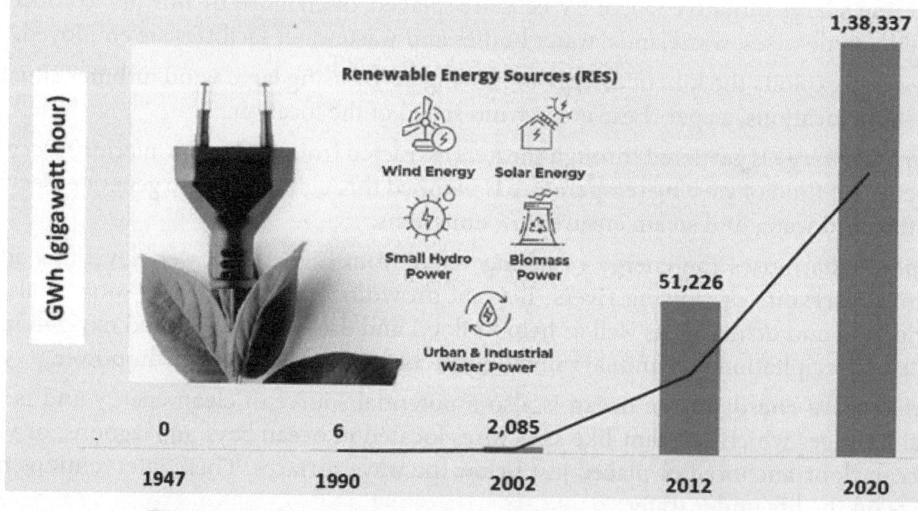

Fig. 3.6: Power Generation from Renewable Energy Sources

(source: Press Information Bureau, GoI)

3.3.2 Non-renewable Sources

Non-renewable energy resources are coal, oil or petroleum, natural gas, and nuclear energy, and are those resources that cannot be replenished easily in consideration of the rate of consumption of the same. Fossil fuels are the largest contributors towards global greenhouse gas and CO_2 emissions, and accounts for over 80% of the world's global energy production. These fossil fuels were formed over millions of years due to the immense heat and pressure within the earth's crust that converted plant and animal matter into coal, oil, and natural gas. The radioactive elements required to generate nuclear energy, usually uranium, comes from mined ore. However, beyond the issue of limited supply, it is the release of high amounts of CO_2 that makes non-renewables an undesired source of energy.

Over 80% of India's requirement is met by coal, oil and biomass, and in spite of having low per capita energy consumption, **India is the 3rd largest global emitter** of CO_2. While the largest domestic source of energy is coal, about 70%, the cumulative domestic fossil fuel production is the lowest among the emerging economies. India relies on crude oil imports as it has only 0.6% and 0.4% of the world's gas and oil reserves respectively. Interestingly however, it is a large net exporter of refined products, amounting to $1.1billion worth to Australia in 2016-17, and has experienced overcapacity in thermal energy. 58.6% of this thermal energy is produced from coal, while the other sources are lignite, diesel, and gas.

3.3.3 New sources and technologies

A promising alternative to traditional fossil fuels is **Hydrogen** as a new source of energy. Hydrogen fuel cells are devices that generate electricity by combining hydrogen and oxygen to produce water and heat as by-products, and can be powered by hydrogen produced from renewable energy sources. They are highly efficient, with an energy conversion rate of up to 60%, compared to around 30% for traditional internal combustion engines. They also produce zero greenhouse gas emissions, which can help to mitigate climate change. However, there are several challenges associated with the adoption of hydrogen fuel cells, such as, the infrastructure required for producing, storing, and distributing hydrogen, and the development of a sustainable and scalable hydrogen production system as the production of hydrogen is currently reliant on fossil fuels. The cost is also a barrier to widespread adoption, although research and development are helping to reduce costs.

Another potent technology for energy generation is **nuclear fusion**, which involves combining of light atomic nuclei to form heavier ones and in the process, releasing vast amounts of potential energy. Unlike nuclear fission, which is currently used in nuclear power plants, nuclear fusion does not produce long-lived radioactive waste and has the potential to provide a virtually unlimited source of clean energy. However, there are several challenges associated with developing nuclear fusion as a practical energy source. Firstly, the high temperature and pressure required to initiate and sustain the fusion reaction are difficult to achieve and maintain for extended periods of time, and the magnetic fields required to contain and control the fusion reaction are complex and difficult to engineer. Secondly, since most nuclear fusion research focuses on using deuterium and tritium–two isotopes of hydrogen, as fuel, the development of a sustainable fuel source is a challenge as tritium is radioactive and must be produced artificially. This requires a supply chain that is currently not fully developed and is also expensive. Furthermore, the development and construction of a commercial-scale fusion reactor would require significant investment and infrastructure.

3.4 WATER RESOURCE MANAGEMENT

Water resource management is a critical to ensure sustainable and equitable access to water resources for human and ecological needs. In recent years, the issue of water scarcity and the need for effective water resource management has become more pressing due to factors such as population growth, climate change, and increasing water demand from various sectors. The World Bank defines Water Resource Management (WRM) as, "*the process of planning, developing, and managing water resources, in terms of both water quantity and quality, across all water uses. It includes the institutions, infrastructure, incentives, and information systems that support and guide water management.*"

The main challenge in water resource management is the need to mitigate the competing demands of various stakeholders, such as, domestic, industrial, agricultural, and environmental. The traditional approach to water resource management has been to focus on supply-side solutions, such as building dams and reservoirs, to increase the availability of water. However, this approach has been criticized for its negative social and environmental impacts, as well as its limited effectiveness in addressing water scarcity in the long term. Recent approaches emphasize the importance of demand-side management strategies, which aim to reduce water consumption and increase efficiency through measures, such as, water conservation, water pricing, and water reuse, which can be effective in reducing water demand, while also providing economic and environmental benefits. Integration of hydrological modelling and remote sensing technologies is also being employed to improve water management decision-making, as these tools provide valuable information on water availability and usage, as well as help to identify areas of water stress and potential risks to water security. However, the role of institutional and governance arrangements in water resource management is pivotal, as effective governance structures and policies are essential for ensuring equitable access to water resources, as well as promoting sustainable and efficient use of water. Participatory and inclusive decision-making processes, as well as the need for greater collaboration and coordination between different stakeholders are much needed to ensure a holistic solution.

3.4.1 WATER, SANITATION AND HYGIENE (WASH)

Ensuring access to water and sanitation is one of the critical SDG (goal 6) as safe water, sanitation and hygiene (WASH) is a basic human need. However, contamination and reclaiming of natural water bodies, overuse of ground water and hindering its replenishment, overall poor management and low investments in water infrastructure are some of the key causes of scarce and unsafe water, and inadequate sanitation.

WASH is a collective term that refers to *a set of interventions aimed at improving access to safe water, adequate sanitation, and proper hygiene practices, and is essential for promoting public health, preventing the spread of disease, and reducing poverty*. **'Water'** stands for access to clean drinking water for household and community use, as well as for agriculture, industry, and other economic activities, **'Sanitation'** implies access to safe and hygienic toilet facilities, as well as the safe disposal of human waste, and **'Hygiene'** refers to practices that promote good health, such as handwashing with soap, safe food handling, and proper menstrual hygiene management. WASH interventions can have significant health and social benefits, particularly for vulnerable populations, especially, women and children. For example, access to safe water and sanitation can reduce the incidence of waterborne diseases, such as, diarrhoea, cholera, and typhoid fever. Improved hygiene practices can also reduce the risk of infectious diseases, improve nutritional outcomes, and promote overall well-being. In turn, it is crucial for achieving several other SDGs , particularly those related to health, gender equality, education and economic growth.

Fig. 3.7: Systemic support required for successful WASH initiatives

(source : www.sdgs.un.org)

WASH implementation in India has been a major focus area for the government, NGOs, and international organizations for many years. Some of the **initiatives taken by the Indian government** to improve WASH include:

1. ***Swachh Bharat Abhiyan*** (Clean India Campaign) is a nationwide campaign launched in 2014 to eliminate open defecation, improve solid waste management, and promote hygiene and cleanliness. The campaign has been successful in increasing access to toilets and reducing open defecation.
2. ***Jal Jeevan Mission***, a flagship program launched in 2019 with the aim of providing tap water connections to every rural household by 2024. The program aims to provide safe and adequate drinking water to all households in rural areas.
3. ***National Rural Drinking Water Programme (NRDW)*** launched in 2009 with the aim of providing safe and adequate drinking water to rural areas. The program focuses on creating sustainable drinking water sources, promoting water conservation, and improving water quality.
4. ***National Urban Sanitation Policy (NUSP)*** launched in 2008 with the aim of promoting sanitation and hygiene in urban areas. The policy focuses on creating sustainable sanitation infrastructure, promoting behaviour change, and improving waste management.

3.4.2 Strategies for water provisioning and management

Water provisioning and management are undertaken through a combination of policies, regulations, and practices that aim to ensure the sustainable use and distribution of water resources.

Some of the **key strategies** employed are as follows:

1. ***Water conservation and efficiency measures***: These include promoting water-efficient technologies and practices, such as low-flow fixtures, drought-resistant landscaping, and water reuse.
2. ***Water pricing and incentives***: Pricing mechanisms and incentives can help to encourage more sustainable water use and conservation practices.
3. ***Investments in water infrastructure***: Investment in water infrastructure such as dams, reservoirs, and treatment plants can improve access to water and sanitation services.
4. ***Integrated water resources management***: This approach aims to manage water resources in a holistic and integrated manner, taking into account the needs of all stakeholders and balancing social, economic, and environmental considerations.
5. ***Water governance and institutional arrangements***: Effective water governance and institutional arrangements are critical for ensuring the equitable distribution and sustainable management of water resources.
6. ***International cooperation***: Global collaboration is critical for addressing transboundary water issues and promoting sustainable water management practices worldwide.

3.5 TELECOMMUNICATION

Telecommunication is defined as the transmission of information – voice, data and multimedia, over electronic media across large distances. It uses various types of transmission technologies, such as, over wire, electromagnetic, radio and optical, and requires infrastructure for radio and television broadcasting, wired and wireless devices, fibre optic cables, satellites, and networks etc. Telecommunication infrastructure is a key driver of economic development, enabling access to information, markets, and resources that can help to spur innovation and growth. The importance of telecommunication infrastructure can be seen in several domains, such as :

1. ***Business and commerce***: Telecommunication infrastructure is essential for conducting business and commerce across distances, facilitating transactions, and enabling access to markets and customers around the world.
2. ***Education and research***: Telecommunication infrastructure is increasingly important in education and research, enabling distance learning, remote collaboration, and access to online resources.
3. ***Public safety and emergency response***: Telecommunication infrastructure is critical for public safety and emergency response, enabling communication between emergency responders and the public during times of crisis.
4. ***Social connectivity and cultural exchange***: Telecommunication infrastructure provides a means for people to connect with each other, regardless of their location, enabling social connectivity and cultural exchange.

3.5.1 Telecom infrastructure

Telecommunication infrastructure refers to the physical networks, equipment, and facilities used to transmit voice, data, and multimedia communication over long distances. There are several types of telecommunication infrastructure, broadly for wired and wireless communication. Wired networks use physical cables, such as copper or fibre optic cables, to transmit data over long distances, as in, traditional landline telephone networks, cable television networks, and high-speed internet networks. Whereas Wireless networks use radio waves to transmit data over the air, without the need for physical cables, such as, cellular networks, satellite networks, and Wi-Fi networks.

However, both require physical facilities and equipment, such as;

- **Telecom towers or cell towers** are structures used to facilitate wireless communication between mobile devices and the telecommunications network. These towers are typically tall structures, ranging from 30 meters to over 100 meters in height, and are equipped with antennas, transmitters, and receivers that enable wireless communication. They are typically located in urban and rural areas, along highways and major roads, and in remote areas where coverage is limited, and maybe freestanding or mounted on existing structures, such as buildings or utility poles. They can also be designed to accommodate multiple carriers, allowing different providers to share the same tower and reduce costs. However, the concern of exposure to electromagnetic radiation and its impact on human health is much debated.
- **Cables** can be installed over ground, underground or under seas. Overground cables are typically strung between poles, buildings, or other structures, and are commonly used in urban and suburban areas, where the cost and complexity of underground installation can be prohibitive. Overground cables are typically made of copper or fibre optic materials, and are designed to withstand weather and environmental conditions. Underground cables are buried beneath the ground either direct-buried or installed in conduits or ducts, and are commonly used in urban areas. While these are more expensive and difficult to install, underground cables are more reliable and secure, however other activities such as road construction or excavations and servicing sewers, etc can lead to damaging of underground cables. Undersea cables are physical cables laid on the ocean floor to connect different continents and regions,

in line with underground but the environmental conditions and installation is extremely challenging.

- **Other Transmission equipment** and devices, i.e., the equipment used to transmit and receive signals, such as, antennas, modems, routers, and switches.
- **Internet exchange points (IXPs)** are physical locations where different internet service providers (ISPs) connect their networks to exchange data. IXPs are essential for enabling the flow of data across different networks and supporting the growth of the internet.
- **Data centers** are facilities used to store, manage, and process large amounts of data, particularly in the areas of cloud computing and internet-based services.

3.5.2 Present and future Challenges in Telecom

Telecommunication infrastructure faces a range of challenges that can impact its effectiveness and ability to support modern communication needs, some of which are as follows:

1. ***Access and Connectivity gaps:*** Despite significant progress in expanding telecommunication infrastructure, there are still areas around the world that lack access to reliable and high-speed internet connectivity, *particularly in rural and remote areas. Addressing these gaps in* connectivity requires significant investment and infrastructure development.
2. ***Cybersecurity threats:*** Telecommunication infrastructure is vulnerable to a range of cybersecurity threats, including hacking, data breaches, and cyber-attacks. Ensuring the security of telecommunication networks and devices is critical to protecting sensitive information and maintaining public safety.
3. ***Cost and affordability:*** Telecommunication infrastructure can be expensive to build and maintain, which can impact its availability and affordability, particularly in developing countries. Ensuring that telecommunication services are accessible and affordable to all is essential for supporting economic development and social connectivity.
4. ***Regulation and policy***: Telecommunication infrastructure is subject to a range of regulatory and policy frameworks that can impact its development and implementation. Ensuring that policies and regulations support the growth and effectiveness of telecommunication infrastructure can be challenging, particularly in rapidly changing technology environments.
5. ***Integrating Emerging technologies:*** The telecommunication industry is constantly evolving, with new technologies and integrating these emerging technologies into existing infrastructure can be challenging, requiring significant investment and expertise.

The India Story

India is the second-largest telecommunications market globally, with a consistently growing subscriber base and broadband subscriptions. As of December 2022, the tele-density reached 84.56%, while broadband subscriptions reached 832.2 million. The total subscriber base stood at 1170.38 million. In the first quarter of FY23, the telecom sector's gross revenue was Rs. 76,408 Crores, i.e., US$ 9.3 billion (IBEF).

In June-September 2022, the total number of internet subscribers reached 850.95 million, with the wireless segment contributing 95.4% of the total telephone subscriptions. Among the different data technologies, 2G accounted for 0.16%, 3G for 1.02%, and 4G for 98.81% of the total wireless data usage. The rise in mobile-phone penetration and declining data costs are expected to bring 500 million new internet users to India in the next five years, creating opportunities for new businesses. Government initiatives such as the BharatNet Project Scheme, Telecom Development Plan, Aspirational District Scheme, and Comprehensive Telecom Development Plan (CTDP) in the North-Eastern Region have led to a significant growth of 200% in rural internet subscriptions from 2015 to 2021. The integration of payments on unified payments interface (UPI) enabled higher and easier usage. Department of Telecommunication (DoT) launched '*Tarang Sanchar*'–a web portal sharing information on mobile towers and EMF Emission Compliances.

In the past four years, there has been over 75% increase in internet coverage, from 251 million users to 446 million, and it is estimated that by 2025, India will require approximately 22 million skilled workers in 5G-centric technologies such as the Internet of Things (IoT), Artificial Intelligence (AI), robotics, and cloud computing.

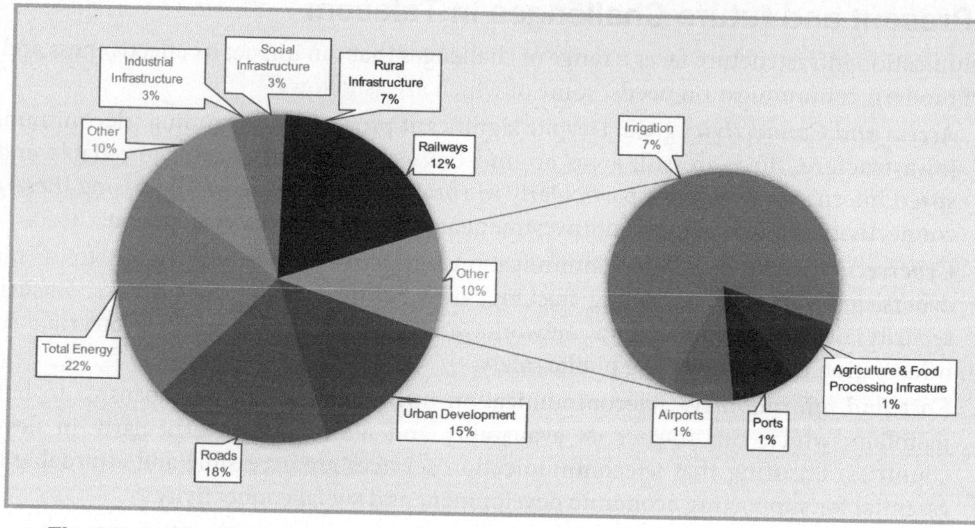

Fig. 3.8: National Infrastructure Pipeline, 2022-25, Sector-wise Fund allocation in %

(source : DoEA, MoF, 2019)

3.6 INFRASTRUCTURE DEVELOPMENT STANDARDS & CODES

The **World Standards cooperation (WSC)**, a collaboration between; International Organization for Standardization (ISO), International Telecommunication Union (ITU), and International Electrotechnical Commission (IEC), sets the global framework for standards for many sectors including infrastructure and forms the backbone for the national standards in India. In the recent Global Quality Infrastructure Index (GQII) 2021, which ranks 184 economies in the world on the basis of the quality infrastructure (QI), **India's national accreditation system** under the Quality Council of India (QCI) has *ranked 5th in the world*, and our overall QI system ranked at the 10th position, with the standardization system (under BIS) at 9th and the metrology system (under NPL-CSIR) at 21st position in the world. The Bureau of Indian Standards (BIS), a statutory body established through legislation through The Bureau of Indian Standards Act 1986, adopts from ISO and IEC standards while the Ministry of Communication and Information Technology have ITU membership. The other standards making bodies of India, apart from BIS are; Telecom Engineering Centre (TEC), Telecommunications Standards Development Society of India (TSDSI), Automotive Research Association of India (ARAI), and Research Designs and Standards organization (RDSO).

In India, the roles and responsibilities for development of infrastructure is defined by the Constitution of India. While the centre is responsible for the critical national-level infrastructures, such as, National highways, Railways, Major ports, Airports and Telecom, the State together with the Municipality and Panchayats are responsible for regulating building construction, water management and supply, urban town planning, other roads and bridges, rural housing and rural electricity. To aid in Public-Private Partnership to see through infrastructural development, in 2007 the Centre set up the **India Infrastructure Project Development Fund (IIPDF) Scheme** aimed at creating appropriate mechanisms, guidelines, advisories, and funding support.

3.6.1 International and National Codes for Civil Engineering and Construction

There are a large number of **Indian Standard (IS) codes** that support the practice of civil engineering and architecture for safe and durable infrastructure construction. These offer guidelines, specifications, and safety prerequisites for construction materials, design parameters, testing techniques, and construction practices, and serve as a reference to guarantee consistency, safety, and excellence in civil engineering projects throughout the country.

Civil engineers refer to these codes for design and analysis of structures, as well as for specifications, methods and code of practice, for e.g., IS : 456; 10262; Sp 23 provides 'codes for designing concrete mixes', while IS : 2386 provides 'methods for tests for aggregate for concrete' and IS : 4925 provides 'specifications for concrete batching plant'. Dedicated list of standards is available for materials and elements, such as, Cement & Concrete, which in turn has codes on cement (IS 269, IS 8041, IS 650), coarse / fine Aggregate (IS 383, IS 2386), Masonry Mortar, Cement Concrete, Curing Compound, etc. Other codes cover Lime and Gypsum, and Doors, Windows and Shutters.

Nationally recognised, some of the **IS codes on civil engineering** are;

Code No.	Description
IS - 4031	Method of physical tests for hydraulic Cement
IS - 650	Specification for Standard sand for testing of Cement
IS - 383	Specification for Coarse and Fine aggregate for use in mass concrete
IS - 515	Specification for natural and manufactured aggregate for use in mass concrete.
IS - 2387	Method of test for aggregates for concrete.
IS - 516	Methods of test for strength of concrete.
IS - 1199	Methods of sampling and analysis of concrete
IS - 3025	Methods of sampling and test (physical and chemical) for water used in industry.
IS - 432	Specification for Mild steel and medium tensile bars and hard drawn steel wire.
IS - 1139	Specification for hot rolled mild steel, medium tensile steel and high yield strength steel deformed bars for concrete reinforcement.
IS - 1566	Specification for plain hard drawn steel wire fabric for concrete reinforcement
IS - 1785	Specification for plain hard drawn steel wire for prestressed concrete.

IS - 1786	Specification for cold twisted steel high strength deformed bars for concrete reinforcement.
IS - 303	Specification for Plywood for general purposes

Internationally recognised, the **ISO ICS 93 codes** are on civil engineering, and covers information on the following;

ISO Code No.	Description
93.010	Civil engineering in general – Construction drawings, see 01.100.30
93.020	Earthworks. Excavations. Foundation construction. Underground works – Including geotechnics. – Earth-moving machinery, see 53.100
93.025	External water conveyance systems – Including buried and above ground installations – Pipelines and its parts for external water conveyance systems, see 23.040.03 Internal water supply systems, see 91.140.60
93.030	External sewage systems – Sewage water disposal and treatment, see 13.060.30 – Pipelines and its parts for external sewage systems, see 23.040.05 Internal drainage systems, see 91.140.80
93.040	Bridge construction
93.060	Tunnel construction
93.080	Road engineering
93.100	Construction of railways – Including the construction of tramways, funicular railways, cableways, rail traffic control equipment and installations, etc. – Rails and railway components, including track, see 45.080 – Equipment for railway/cableway construction and maintenance, see 45.120
93.110	Construction of ropeways – Ropeway equipment, see 45.100 – Equipment for ropeway construction and maintenance, see 45.120
93.120	Construction of airports – Including air transport control equipment and installations
93.140	Construction of waterways, ports and dykes – Including river embankments, water transport control equipment and installations, etc.
93.160	Hydraulic construction – Hydraulic energy equipment, see 27.140

In addition, the **National Building Code (NBC),** discussed in detail in Unit 5, a model code for adoption of across the nation and contains administrative regulations, development control rules and general building requirements; fire safety requirements; stipulations regarding materials, structural design and construction (including safety); construction management practices and safety, building and plumbing services; approach to sustainability; and asset and facility management is referred to during design and detailing construction working drawings .

Eventually the quality of design and execution is the onus of the Civil engineer and Architect. Thus, to aid in other complimentary areas, Niti Ayog has compiled the '**Indian Infrastructure Body of Knowledge** – A technical Baseline to the Practice of Program and Project Management in India' under the National Program and Project Management Policy Framework with the intention of "*laying down a plan of action and advocating short-term and long-term strategies for improving Program and Project Management practices in India, as well as align with the global best practices*" (CEO, Niti Ayog)

3.7 INNOVATIONS AND METHODOLOGIES FOR SUSTAINABILITY

Sustainable Infrastructure, be it built, natural or hybrid, are systems that are "*planned, designed, constructed, operated and decommissioned in a manner that ensures economic and financial, social, environmental, including climate resilience, and institutional sustainability over the entire infrastructure life cycle*" (UNEP). The Organisation for Economic Co-operation and Development (OECD) estimates that an annual investment of USD 6.9 trillion is needed for infrastructure to meet development goals and create a low carbon, climate resilient future by 2050. OECD's ***Strategic Policies for Sustainable Infrastructure*** identifies various themes that require attention, such as, Low-carbon transition, Technology and Innovation, Inclusiveness and Accessibility, etc.

Presently the OECD is developing a toolkit on quality infrastructure investment for policymakers and practitioners, based on the ***G20 Principles for Quality Infrastructure Investment (QII)***, developed under the Japanese G20 Presidency, that stated, "*quality infrastructure investment contributes to maximizing the positive impact of infrastructure to achieve sustainable growth and development, raising economic efficiency in view of life cycle costs, integrating environmental and social considerations in infrastructure, building resilience against natural disasters, and strengthening infrastructure governance.*"

The UN Environment Assembly (UNEA) Members States in March 2023 adopted a resolution on '***Sustainable and Resilient Infrastructure***' encouraging them to;

- provide opportunities for engaging relevant stakeholders,
- promote investment in sustainable and resilient infrastructure, natural infrastructure and nature-based solutions,
- cooperate internationally to strengthen frameworks, including for financing, and
- implement the '*International Good Practice Principles for Sustainable Infrastructure*', which in turn has the following guiding principles ;
    1. Strategic Planning
    2. Responsive, Resilient, And Flexible Service Provision
    3. Comprehensive Life Cycle Assessment of Sustainability
    4. Avoiding Environmental Impacts and Investing In Nature
    5. Resource Efficiency and Circularity
    6. Equity, Inclusiveness, And Empowerment
    7. Enhancing Economic Benefits
    8. Fiscal Sustainability and Innovative Financing
    9. Transparent, Inclusive, And Participatory Decision-Making

10. Evidence-Based Decision-Making

There are several **innovations and methodologies** that can be employed to ensure the sustainability of infrastructure development, such as :

1. ***Green infrastructure*** is a concept of incorporating the importance of Environment and considering the impact of decisions on it while developing infrastructure strategies. It involves the use of natural systems and materials to provide sustainable solutions, for e.g., green roofs, permeable pavements, and rain gardens, which can help to reduce stormwater runoff and mitigate the urban heat island effect.
2. ***Integrated Water Resources Management (IWRM)*** is a process that promotes the "*coordinated development and management of water, land and related resources in order to maximise economic and social welfare in an equitable manner without compromising the sustainability of vital ecosystems*" (UNEP).
3. ***Circular Economy*** or circularity is an economic model of production and consumption that seeks to eliminate waste and promote the sustainable use of resources. In the context of infrastructure development, it can be used to design infrastructure projects that prioritize the use of renewable materials, reduce waste, and promote the reuse and recycling of materials.
4. ***Tool and methods for reducing Environmental Impact***, such as, Environmental Impact Assessment (EIA) is a methodology used "*to identify the environmental, social and economic impacts of a project prior to decision-making*, Life-Cycle Assessment (LCA) is a tool used to "*evaluate the environmental impact of infrastructure projects throughout their entire life cycle, from raw material extraction to disposal*", and Building Information Modelling (BIM) is a digital modelling technology that enables the creation of detailed 3D models of infrastructure projects, to help improve project efficiency, reduce waste, and optimize the use of materials and resources.

Infrastructure design and development is crucial for a Nation's growth and civil engineers play the most pivotal role in ensuring quality, safety, innovation, and sustainability of the same, bearing great societal and global impact.

UNIT SUMMARY

This unit focusses on present and future projection of various facets of Infrastructure development by discussing concepts such as 'Smart City' and further delves into understanding critical infrastructure, such as, Transportation, Energy, Water resources management and Telecommunication. It also offers an overview on the Standards and Codes relevant for infrastructure and construction industry, and culminates with a discussion on present day initiatives, innovations and methodologies employed in this sector to ensure sustainability.

EXERCISES

I. Multiple Choice Questions

Q.1 Which of the following are 'critical infrastructure' as per India ?

a. Disaster Management b. Aviation

c. Cyber security d. all of the above

Q.2 What is the population of a 'Megacity' ?

a. 1 to 5 million b. over 10 million

c. less than 10 million
d. 1 million

Q.3 Which ancient Indian University town, now in present day Afghanistan, was connected to Prayagraj by the Great Road built by Mauryan Empire?

a. Bhagalpur
b. Gandhar
c. Takshashila
d. Benaras

Q.4 What is the world's longest road tunnel?

a. the Lærdalstunnelen
b. Channel Tunnel
c. Yamate Tunnel
d. Gotthard Base Tunnel

Q.5 Which of the following renewable energy source may be argued to not be a 'clean energy' source?

a. Solar
b. Biomass
c. Geothermal
d. Wind

Answers of Multiple Choice Questions				
3.1 **d**	3.2 **b**	3.3 **c**	3.4 **a**	3.5 **b**

II. Short and Long Answer Type Questions

Q.6 What is a 'Smart City'? Briefly explains the characteristics of a Smart City.

Q.7 What are some of the potential new sources of energy? What are the associated challenges and impacts?

Q.8 Define WASH? What are the various initiatives by Govt. of India to promote WASH?

Q.9 What are the various types of physical facilities and equipment required as part of Telecommunication infrastructure? Discuss the challenges that impact Telecommunication infrastructure?

Q.10 What is the importance of Infrastructure development standards and codes? Illustrate with examples how this supports the profession of civil engineering.

KNOW MORE

About Futuristic Cities

https://www.dezeen.com/2022/08/01/futuristic-cities-planned-architecture-masterplanning-urban-design/#

About Top 15 navigable canals

https://marine-digital.com/article_top_15_canals

About Hyperloop concept

https://www.tesla.com/sites/default/files/blog_images/hyperloop-alpha.pdf

About standards

https://sesei.eu/indian-standardization/national-standardization-bodies/

Civil Engineering Codes

http://www.civilology.com/is-codes-list-in-civil-engineering/

InBoK , Niti Ayog

https://www.niti.gov.in/sites/default/files/2020-11/The-Indian-Infrastructure-Body-of-Knowledge-Rev-C.pdf

India Infrastructure Project Development Fund

https://mohua.gov.in/upload/uploadfiles/files/Guideline_Scheme_IIPDF.pdf

International Good practice Principles for Sustainable Infrastructure

https://wedocs.unep.org/bitstream/handle/20.500.11822/39811/infrastructure_practices2.pdf?sequence =1&isAllowed=y

REFERENCES AND SUGGESTED READINGS

Books and Publications

- [Book] Airport system development. (1984). United States: DIANE Publishing.
- Aschauer, D. A. (1990). *Why is infrastructure*. Industry Week, 21-50.
- Bordoloi, T., and Acharya, S. (2023). *A systematic literature review on Sustainable Smart City Planning to strategize design for North-east India*. In Proceedings of International Conference on Research into Design. Springer, Singapore.
- Li, D., Ma, J., Cheng, T., van Genderen, J. L., & Shao, Z. (2018). *Challenges and opportunities for the development of MEGACITIES*. International Journal of Digital Earth.
- Prud'homme, R. (2004). *Infrastructure and development* (pp. 3-5). Washington, DC: World Bank.
- Sapatnekar,S., Patnaik,I., and Kishore,K. (2018) *Regulating Infrastructure Development in India, National Institute of Public Finance and Policy, Working Paper No. 230.* < https://macrofinance.nipfp.org.in/PDF/PatnaikSapatnekarKishore- regulating_infrastructure_development_in_india.pdf >
- Wenzel, F., Bendimerad, F., & Sinha, R. (2007). *Megacities–megarisks*. Natural Hazards, 42, 481- 491.

Web sources

- https://www.eea.europa.eu/themes/biodiversity/an-introduction-to-habitats
- https://habitat3.org/documents-and-archive/new-urban-agenda/
- https://www.un.org/en/development/desa/population/publications/pdf/urbanization/the_worlds_cities_in_2018_data_booklet.pdf
- https://www.ibef.org/government-schemes/smart-cities-mission
- https://pib.gov.in/PressReleaseIframePage.aspx?PRID=1888480
- https://irtpms.in/site/wp-content/uploads/2017/09/civil-engg.pdf
- https://www.ief.org/programs/infrastructure https://digitalsky.dgca.gov.in/home
- https://www.un.org/en/climatechange/what-is-renewable-energy

04 ENVIRONMENT

UNIT SPECIFICS

Through this unit we have discussed the following aspects:

- ***Environment, Global Warming and Climate Change***
  - *Global warming phenomena and GHG emissions,*
  - *Pollution Mitigation – Measures and Approaches,*
  - *Non-stationarity*
- ***Environmental Metrics and Monitoring***
  - *Environmental Monitoring*
  - *Global Climate Indicators and Essential Climate Variable (ECV)*
  - *Environmental Performance Index Indicators*
  - *Other Sustainability measures*
- ***Innovations and methodologies for ensuring Sustainability***
  - *Environmental Impact Assessment (EIA)*
  - *Life cycle Assessment (LCA)*
  - *Strategic Environmental Assessment (SEA)*

Besides giving a large number of multiple choice questions as well as questions of short and long answer types marked in two categories following lower and higher order of Bloom's taxonomy, a list of references and suggested readings are given in the unit so that one can go through them for practice.

There is a "Know More" section, which has been carefully designed so that the supplementary information provided in this part becomes beneficial for the users of the book. It is important to note that for getting more information on various topics of interest some QR codes have been provided which can be scanned for relevant supportive knowledge This section mainly highlights applications of the subject matter for our day-to-day real life or/and industrial applications on variety of aspects, case study related to environmental, sustainability, social and ethical issues whichever applicable, and finally inquisitiveness and curiosity topics of the unit.

RATIONALE

This foundational unit on environment and the impact associated with it, offers understanding of the various views on the state of the environment at present; and provides knowledge on mitigation strategies and policies, metrices and indicators, and most importantly, methodologies and tools that are applied to assess environmental impact, to empower the civil engineer in practice.

UNIT OUTCOMES

List of outcomes of this unit is as follows:

U4-O1: *Understanding of Global warming phenomena, Climate change and Pollution Mitigation*

U4-O2: *Knowledge of Environmental Monitoring & Metrics*

U4-O3: *Knowledge on innovations and methodologies for ensuring Environmental Sustainability*

Unit-4 Outcomes	EXPECTED MAPPING WITH COURSE OUTCOMES (1- Weak Correlation; 2- Medium correlation; 3- Strong Correlation)						
	CO-1	CO-2	CO-3	CO-4	CO-5	CO-6	CO-7
U4-01	2	2	3	1	2	3	2
U4-02	3	3	3	1	3	3	2
U4-03	3	3	3	1	3	3	2

"A clean, healthy and sustainable environment is a human right" – was adopted as a landmark resolution by The United Nations Human Rights Council in October 2021. India's Environment (Protection) Act, 1986, Section 2, defines Environment as follows; "*environment includes water, air and land and the inter-relationship which exists among and between water, air and land, and human beings, other living creatures, plants, micro-organism and property*". This Act was instituted by the Govt. of India following the nation's participation in the UN Conference on the Human Environment, Stockholm in June 1972, to prioritise the protection and improvement of environment, and prevent hazards to human beings, other living creatures, plants and property. Climate change and Global warming are commonly used today, almost interchangeably, to imply hazardous impacts on all life and argue the need for Environmental consideration and sustainable consumption.

4.1 ENVIRONMENT, GLOBAL WARMING AND CLIMATE CHANGE

The term '**global warming**' refers to the long-term heating of the Earth's surface observed since the beginning of the Industrial Era due to 'greenhouse effect' caused by human activities, primarily attributed to burning of fossil fuels and industrial processes, and deforestation which leads to significant greenhouse gas emissions. This term should not be used interchangeably with Climate change, as the latter refers to the long-term change in the average weather patterns – temperature, precipitation, wind and tidal patterns, and is not limited to the adverse effects of human activities alone. Natural causes such as, volcanic activity, cyclical ocean patterns, orbital changes, may also contribute to climate change. **Climate change** has far-reaching consequences beyond temperature increase, as it affects ecosystems, agriculture, water availability, human health, and socio-economic systems. Key indicators of Climate change are; frequency and severity changes in extreme weather such as hurricanes, heatwaves, wildfires, droughts, floods, and precipitation; ice loss at Earth's poles and in mountain glaciers; rising sea levels; cloud and vegetation cover changes, as well as the global land and ocean temperature increases. It maybe oversimplified to state that global warming leads to climate change.

4.1.1 Global warming phenomena and GHG emissions

The phenomenon of global warming occurs when greenhouse gases or GHG, such as, carbon-dioxide (CO2), chlorofluorocarbons, water vapour, methane (CH4), and nitrous oxide (N2O), which are heat-trapping pollutants create a layer in the Earth's atmosphere which disallows the solar radiation emitted by the earth's surface to escape and instead absorbs the heat, thereby increasing the surface temperature. This **greenhouse effect** is a natural process that is essential for maintaining Earth's habitable temperature. However, human activities, especially the burning of fossil fuels (coal, oil, and natural gas), deforestation, and industrial processes, have significantly increased greenhouse gas emissions over the past century. The burning of fossil fuels for electricity generation, transportation, and industrial processes is the primary source of CO2 emissions. Agricultural practices, such as livestock production and rice cultivation, contribute to CH4 and N2O emissions. As the concentration of greenhouse gases increases, more heat is trapped within the Earth's atmosphere, causing a rise in average global temperatures, in turn, leading to the melting of polar ice caps and rising sea-level. Increased CO2 absorption by the oceans can lead to ocean acidification, which have detrimental effects on marine life, including coral reefs and shell-forming organisms. The changing temperatures can intensify extreme weather events, such as, hurricanes, heatwaves, droughts, and alter precipitation, which disrupts the ecosystem and affects plant and animal species' distribution and migration patterns.

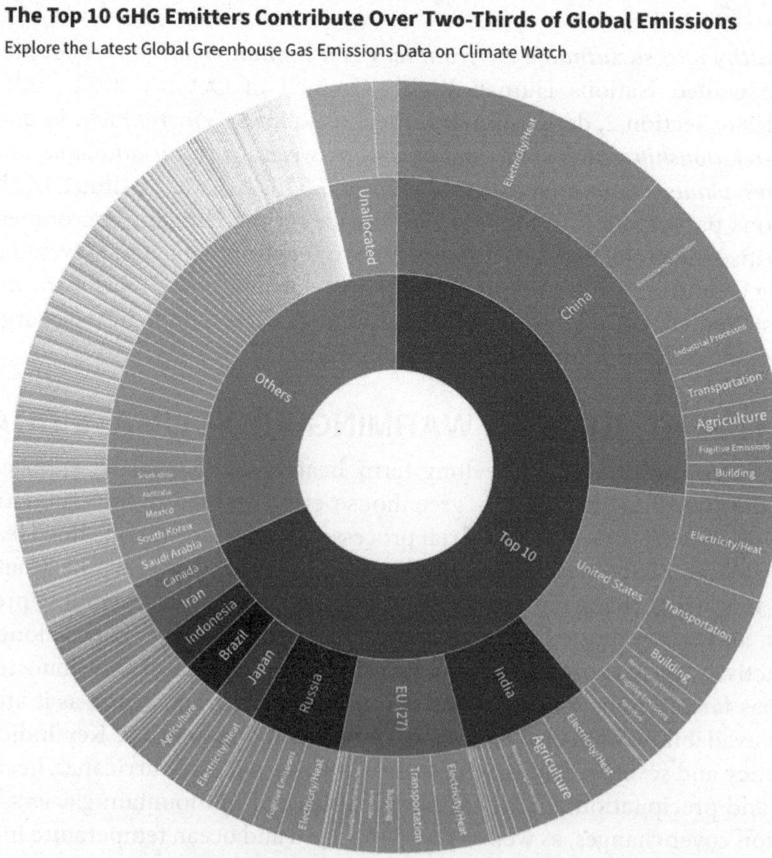

Fig. 4.1: Global GHG Emissions, 2019, by Country and Sector

source:www.climatewatchdata.org

The biggest GHG emitting nations are China, USA, Russia, and India. The ***top 10 emitters*** (country-wise) account for over **two-thirds of annual GHG emissions**, together accounting for over 50% of the global population and 75% of the world's GDP. GHG emissions are usually measured in *CO_2 equivalent*, as CO_2 alone accounts to almost 76-78% of global GHG emissions, however the jury is still out on the strength of CO_2 to cause warming and its factual effects directly on global temperature rise, as a certain amount is essential in the atmosphere to support all life on this planet, as it is an important nutrient for trees, plants and crops. Other gases need to be converted to this unified unit by multiplying its emission to its Global Warming Potential (GWP), which is a measure of how much energy the emissions of 1 ton of a gas will absorb over a given period of time, relative to the emissions of 1 ton of carbon dioxide (CO_2). GWP acknowledges the fact that many gases are more impactful at warming Earth than CO_2, per unit mass.

The UN's Intergovernmental Panel on Climate Change (IPCC) forewarns that to limit global warming to 1.5°C, GHG emissions must peak before 2025 at the latest and decline 43% by 2030, as they indicate that crossing the 1.5°C threshold risks unleashing severe climate change impacts, including more frequent and severe droughts, heatwaves and rainfall. A landmark, multilateral climate change process – the **Paris Agreement**, binds 196 parties (nations) together to combat climate change and adapt to its effects, was adopted at the UN Climate Change Conference (COP21) in December 2015. Each Party has set 2030 targets in their Nationally Determined Contributions (NDCs) to align with the Paris Agreement temperature goal and has formulated *'long-term low greenhouse gas emission development strategies'* (LT-LEDS) that provides long-term horizon to the NDCs. Under an *'enhanced transparency framework'* (ETF), due to start in 2024, countries will be able to report transparently on actions taken and progress in climate change mitigation, adaptation measures and support provided or received, and feed the information into the **Global stocktake** which will assess the collective progress towards the long-term climate goals. Presently, this is undertaken by various organisations, such as, US Environment Protection Agency, European Environment Agency, International Energy Agency, or World Data Lab, who launched the *World Emissions Clock* at COP27.

The EPA tracks the total emissions in the United States by publishing the *Inventory of U.S. Greenhouse Gas Emissions and Sinks*, and their 2021 report notes that the primary sources of GHG emissions by economic sector are ; Transportation (28%), Power/electricity generation (25%), Industry (23%), Commercial and Residential (13%), Agriculture (10%) and Land use and Forestry (12%). Globally, Power and Energy sector is the largest contributor towards GHG emissions, and transportation and buildings have the major most activities that include both direct emissions from fossil fuel combustion, as well as indirect emissions such as use of electricity. The three sectors that stand out as the fastest-growing sources of GHG emissions since 1990 are–Industrial processes (grew by 203%); Electricity and heating, a subsector of energy (by 84%); and Transportation, also a subsector of energy (by 78%) as noted by EPA (Climate watch, 2021).

4.1.2 Pollution Mitigation

Mitigating global warming requires collective efforts and a comprehensive approach involving various sectors and stakeholders, by leveraging and building on existing mitigation measures and by incorporating sustainable approaches to ensure long-term resolution.

Emission and Pollution Mitigation Measures

Some **key strategies** for mitigating GHG emissions and pollution are as follows:

1. ***Transitioning to Renewable Energy:*** Renewable sources, such as, solar, wind, hydro, and geothermal power from fossil fuels reduces greenhouse gas emissions associated with electricity generation and decreases reliance on carbon-intensive energy sources. Another potential source for generating electricity is from nuclear energy rather than the combustion of fossil fuels.
2. ***Energy Efficiency:*** Largely, it can be achieved by; "*increasing the efficiency of existing fossil fuel-fired power plants by using advanced technologies, substituting less carbon- intensive fuels, and shifting generation from higher-emitting to lower-emitting power plants*" and by "*reducing electricity use and peak demand by increasing energy efficiency and conservation in homes, businesses, and industry.*" (Contribution of Working Group III to the Sixth Assessment Report of the Intergovernmental Panel on Climate Change, chapter 6, 2021). Improving energy efficiency in transportation, buildings, and industrial processes can significantly reduce greenhouse gas emissions. This can be achieved through measures, like; switching to alternate fuels like biofuels, hydrogen, and electricity from renewable sources, or fossil fuels that are less CO_2-intensive; improving fuel efficiency and employing operating practices that minimize fuel use across all sectors. Encouraging the use of active transportation modes, like, walking and cycling, alongside improving public transportation systems and promoting it, reducing travel demand, especially office commute, through policy and urban planning to address long daily commutes will help reduce emissions from the transportation sector. In the building and allied services sectors, measure such as, improving insulation and building envelope design; incorporating efficiency strategies in heating, cooling, ventilation, and refrigeration systems; using LED lighting and passive heating and lighting to take advantage of natural sunlight; using energy-efficient appliances and electronics; incorporating energy efficiency practices into water and wastewater plants, solid waste management, and urban processes, etc. have great potential to curb emissions.
3. ***Forest Conservation and Reforestation:*** Forests act as carbon sinks, absorbing CO_2 from the atmosphere, therefore, protecting existing forests, preventing deforestation, and implementing reforestation and afforestation projects can help sequester carbon and mitigate global warming. Carbon Capture and Sequestration (CCS) where CO_2 is captured as a by-product of fossil fuel combustion, before it enters the atmosphere, and transporting to inject it deep underground at carefully selected and suitable subsurface geologic formation to be securely stored is also a potent mitigation strategy.
4. ***Sustainable Agriculture and Land Use***: Implementing sustainable agricultural practices that reduce emissions from farming activities by emulating natural processes, promoting agroforestry, and minimizing land-use change and deforestation in favour of carbon-rich ecosystems can contribute to global warming mitigation. "*It uses up to 56 per cent less energy per unit of crops produced, creates 64 per cent fewer greenhouse gas emissions per hectare and supports greater levels of biodiversity than conventional farming.*" (UNEP, Trade and Environment Briefings: Sustainable Agriculture, 2012).

Circular Economy Approach to mitigate

The EPA endorses the "**Circular economy for all**" approach (Circular Economy systems diagram, 2019, Ellen MacArthur Foundation) to reduce waste and hazardous materials, and reuse critical minerals in manufacturing, as natural resource extraction and processing contributes to about half of all global GHG emissions (United Nations' International Resource Panel). The UN Conference on Trade and Development (UNCTAD) elaborates that circular economy is an economy where, "*all forms of waste, such as clothes, scrap metal and obsolete electronics, are returned to the economy or used more efficiently*" and "*entails markets that give incentives to reusing products, rather than scrapping them and then extracting new resources*".

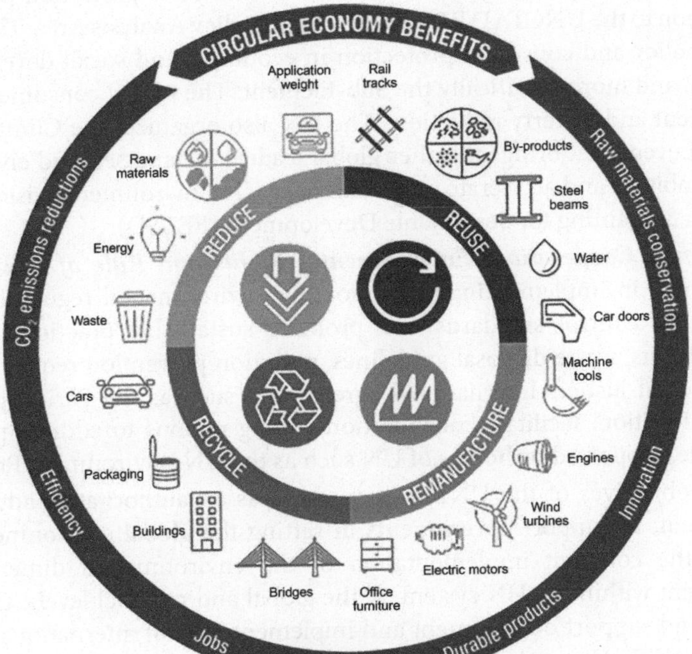

Fig. 4.2: Circular Economy, UNCTAD

The **policies and strategies** developed in line with this approach for mitigating global warming are :

5. ***Waste management and Industrial Best Practices:*** Emphasizing the principles of 4R synonymous to the concept of circular economy, i.e., reduce, reuse, recycle and remanufacture, can help reduce emissions associated with, raw material extraction through conservation, across the entire life of the product from production till disposal. Encouraging industries to adopt cleaner production techniques, such as using less toxic materials, implementing energy-efficient processes, and optimizing resource usage, can help reduce pollution emissions and minimize environmental impacts.

6. ***Sustainable consumption and production:*** Individuals can contribute to global warming mitigation by making sustainable choices in their daily lives, which includes conserving energy, reducing food waste, supporting eco-friendly products, and advocating for sustainable practices. SDG 12 is about ensuring sustainable consumption and production patterns, as present unsustainable behaviour is the root cause of pollution, biodiversity loss and climate change. This is also key to sustain the livelihoods of current and future generations. The UN further adds that "*Governments and all citizens should work together to improve resource efficiency, reduce waste and pollution, and shape a new circular economy.*" Moreover, raising

awareness about pollution-related issues, educating the public on the importance of pollution prevention, and promoting environmentally friendly behaviours can lead to positive changes in individual and collective actions. "*Around 90 per cent of countries report that education for sustainable development and global citizenship education are at least partially mainstreamed in national education laws and policies, curricula, teacher education or student assessments in primary and secondary school.*" (SDG Report 2022). The UNCTAD is also working on research into 'incorporating sustainability into consumer protection policy aims to make a contribution to the UNCTAD RPP Research and Policy Analysis Area: The role of competition law and policy and consumer protection in economic and social development and poverty reduction, and more specifically the Sub-Element: The role of consumer protection in social development and poverty reduction.' The UN also organizes the Climate Action Summit, a high-level event that brings together global leaders, businesses, and civil society to enhance climate ambition and accelerate action, and has UN Environment's vision for Environmental Education & Training for Sustainable Development (EETSD).

7. ***International Cooperation, Environmental Rights and Rule of Law:*** Governments play a crucial role in implementing and enforcing environmental regulations and policies that set pollution control standards and promote sustainable practices, which may include emission limits, waste disposal guidelines, pollution prevention requirements, and ensuring environmental justice. International agreements, such as the Paris Agreement for climate change mitigation, facilitate collaboration among nations to address pollution issues on a larger scale, supported by bodies of UN such as the UN Environment Program (UNEP).

8. The core objectives of the UNEP are to serve as an authoritative advocate for the global environment, to support governments in setting the global environmental agenda, and to promote the coherent implementation of the environmental dimension of sustainable development within the UN system, at the global and regional levels. One such means is to promote and support development and implementation of international environmental law. In 2013, UNEP's Governing Body adopted Decision 27/9, on *Advancing Justice, Governance and Law for Environmental Sustainability*, to establish the term '**Environmental rule of law**'. The Environmental rule of law is defined as "*the rule of law as a principle of governance in which all persons, institutions and entities, public and private, including the State itself, are accountable to laws that are publicly promulgated, equally enforced and independently adjudicated, and which are consistent with international human rights norms and standards.*" (Report on the Rule of Law and Transitional Justice in Conflict and Post-Conflict Societies). International environmental law is "*a branch of public international law–a body of law created by States for States to govern problems that arise between States. It is concerned with the attempt to control pollution and the depletion of natural resources within a framework of sustainable development.*" (UNEP, 2017).

9. The UNEP also hosts the Climate Technology Centre and Network (CTCN), the implementation arm of the Technology Mechanism of the United Nations Framework Convention on Climate Change. The Centre promotes "*the accelerated transfer of environmentally sound technologies for low carbon and climate resilient development at the request of developing countries. The CTC provides technology solutions, capacity building and advice on policy, legal and regulatory frameworks tailored to the needs of individual countries by harnessing the expertise of a global network of technology companies and institutions*". Addressing global warming requires international cooperation and the implementation of effective policies at the national and

international levels. This includes setting ambitious emission reduction targets, implementing carbon pricing mechanisms, and fostering collaborations to support technology transfer and capacity building in developing countries.

4.1.3 Non-stationarity

Statistical analysis of the difference between mean CO_2 concentration (ppm) in the 1980s and mean CO_2 concentration in the 2000s, captured by Mauna Loa, Hawaii, which has the longest continuous record of direct atmospheric CO_2 measurements, showed that there has been a significant change in mean maximum temperature from decade to decade, mean maximum temperature from decade to decade for spring, mean maximum temperature from season to season, and further established that there is statistical evidence to claim that climate variables are changing over time (causeweb.org). The climate system is constantly changing across time, and these changes are usually associated with the temporal (time related) variation of the statistical properties of climatic variables, such as temperature, precipitation, wind speed, relative humidity, volcanic activity, seasonal change, and solar radiation levels. Thus, statistically it poses non-stationarity, i.e., the status of a time series whose statistical properties are changing through time, and is opposite of a stationary time series which has statistical properties or moments (e.g., mean and variance) that do not vary in time. Beuno de Mesquita, et al. (2020) define **non-stationarity** as "*a change in the relationship, either in direction or magnitude (from a significant relationship to no relationship and vice versa) between variables over time*". They further add that "*such non-stationarity is often combined with the additional hurdle of needing multivariate statistics*" as the changes in climate have become exceedingly erratic such that previous thought to be coupled climatic characteristics, such as, temperature and drought, are no longer behaving in tandem but now need to be considered as potentially independent.

4.2 ENVIRONMENTAL MONITORING & METRICS

Environmental impact is succinctly described as "*changes in the natural or built environment, resulting directly from an activity, that can have adverse effects on the air, land, water, fish, and wildlife or the inhabitants of the ecosystem. Pollution, contamination, or destruction that occurs as a consequence of an action, that can have short-term or long-term ramifications is considered an environmental impact.*" (Abdallah, 2017). These impacts require a system or standard of measurement, i.e., metric, that indicates the state of a system and measures the its behaviour. Therefore, the term '**indicator**' is commonly used to refer to these metrics for ascertaining the content and performance of a system, corresponding to any of the three dimensions of sustainability–ecological metrics, economic metrics, and sociological metrics (Sikdar,2003). However, holistic understanding, measuring and monitoring can only be achieved at the intersection of these three dimensions, as most systems are complex and are inadequately represented. Therefore, '**true sustainability metrics**' are 3-Dimensional and are classified as follows;

Group 1 (1-D): economic, ecological, and sociological indicators

Group 2 (2-D): socio-economic, eco-efficiency, and socio-ecological indicators

Group 3 (3-D): sustainability indicators

4.2.1 Environmental Monitoring

A large number of sustainability indictors have been thematically classified with respect to the 17 SDGs and discussed in earlier Unit, and there are various means for each to be measured and monitored. *The systematic collection, analysis, and interpretation of data on various aspects of the environment, involving measuring, observing, and assessing environmental indicators to understand the current state, changes, and trends in natural systems and human-induced activities* is referred to as **Environmental monitoring**. The data collected through environmental monitoring helps in assessing the effectiveness of environmental policies, identifying potential risks and hazards, and informing decision-making processes related to resource management, conservation, and pollution control. It encompasses a wide range of parameters, including air quality, water quality, soil health, biodiversity, climate variables, noise levels, and pollution levels, and involves the use of various monitoring techniques, such as sampling, remote sensing, sensor networks, and data analysis tools, to gather accurate and reliable information for environmental assessment and management.

Below enlisted are some common types of **environmental monitoring and the tools/methods** used for each:

1. ***Air Quality Monitoring:***
   - Tools: Air quality monitors (with CO_2 and O_2 sensors), air sampling equipment, Temperature and Humidity monitors, particulate matter (PM) samplers, gas analysers.
   - Methods: Continuous monitoring stations, passive samplers, mobile monitoring, remote sensing, and modelling techniques.
2. ***Water Quality Monitoring:***
   - Tools: Water quality sensors, water samplers, pH meters, turbidity meters, dissolved oxygen meters, Conductivity probes.
   - Methods: Grab sampling, automated water quality monitoring stations, remote sensing, and satellite imagery.
3. ***Soil Monitoring:***
   - Tools: Soil sampling equipment, soil moisture sensors, pH and nutrient analysers.
   - Methods: Soil sampling and analysis, soil moisture monitoring, soil fertility testing, spectrometry (to measure contamination in soils), remote sensing (to monitor salinity in soils).
4. ***Biodiversity Monitoring:***
   - Tools: Camera traps, acoustic sensors, GPS devices, binoculars, species identification guides.
   - Methods: Field surveys, transect sampling, mark and recapture techniques, bio- acoustic monitoring, remote sensing.
5. ***Noise Monitoring:***
   - Tools: Sound level meters, noise dosimeters, acoustic monitoring systems.
   - Methods: Point measurements, continuous monitoring stations, noise mapping, community noise surveys.
6. ***Pollution & Waste Monitoring:***
   - Tools: Gas analysers, spectrometers, emission measurement devices, pollutant- specific sensors.
   - Methods: Source emission sampling, ambient air monitoring, industrial effluent monitoring, pollution source tracking, production and consumption patterns.
7. ***Radiation Monitoring:***
   - Tools: Radiation detectors, dosimeters, Geiger-Muller counters.
   - Methods: Radiation monitoring stations, personal monitoring, environmental sampling, remote sensing.

8. ***Climate Monitoring:***
   - Tools: Weather stations, temperature loggers, rainfall gauges, anemometers, radiometers.
   - Methods: Meteorological observations, climate stations, satellite-based measurements, climate modelling.

4.2.2 Global Climate Indicators and Essential Climate Variable (ECV)

The *World Meteorological Organization (WMO)* **State of the Global Climate** uses seven Global Climate Indicators–Surface Temperature and Ocean Heat (Temperature and Energy); Atmospheric CO2 (Atmospheric composition); Ocean Acidification and Sea-level (Ocean and Water); Glaciers and Arctic and Antarctic Sea Ice Extent (Cryosphere) —to monitor the domains most relevant to climate change, "*without reducing climate change to only temperature*". In addition to the seven headline indicators, a supplementary set of subsidiary indicators is complimentarily available to offer information and contribute to a more comprehensive and detailed depiction of the evolving trends in their respective areas. These indicators are derived from analysis and interpretation of fundamental measurements or parameters for characterizing and understanding the Earth's climate system.

An **Essential Climate Variables (ECV)** is defined by WMO as "*a physical, chemical or biological variable or a group of linked variables that critically contributes to the characterization of Earth' s climate*". These are compiled into datasets to, provide empirical evidence that helps understand and predict the evolution of climate, and assess risks and enable attribution of climate events to underlying causes. EVC datasets *(refer Fig. 3)* further guide mitigation and adaptation measures, and supports the United Nations Framework Convention on Climate Change (UNFCCC) and Intergovernmental Panel on Climate Change (IPCC) efforts for assessment and creating climate services. Assessments provide policymakers and the public with synthesized information on the state of the climate system, projected future changes, and associated impacts, by quantifying the observed changes and attributing them to human activities. Climate services provide tailored climate information and predictions to various sectors, such as, agriculture, water resources, energy, and disaster management, and helps develop climate models, downscaling techniques, and decision support tools.

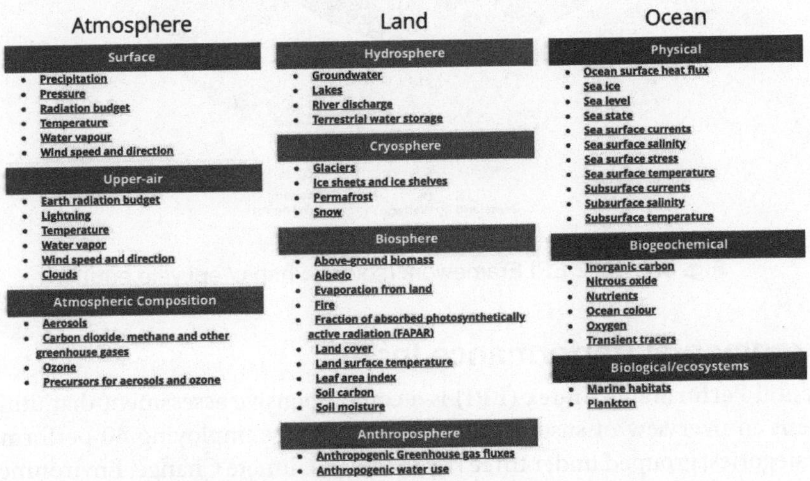

Fig. 4.3: Essential Climate Variables, WMO

source : https://gcos.wmo.int/en/essential-climate-variables

The National Centers for Environmental Information (NCEI) of the U.S. National Oceanic and Atmospheric Administration (NOAA) and the U.S. GCOS Program at NCEI, maintains the Global Observing Systems Information Center (GOSIC) which provides further background, definitions, requirements, network information and data sources for ECVs. These variables are monitored as per the GCOS Climate Monitoring Principles (*agreed by the Committee on Earth Observation Satellites (CEOS) and adopted by the Congress of the World Meteorological Organization (WMO) and Conference of the Parties (COP-9, 2003) to the UNFCCC*), which provide a standardized framework for monitoring and measuring key climate parameters consistently across different regions and over extended periods. The ECVs also help in validating and calibrating climate models, ensuring they accurately represent the observed climate variability and trends.

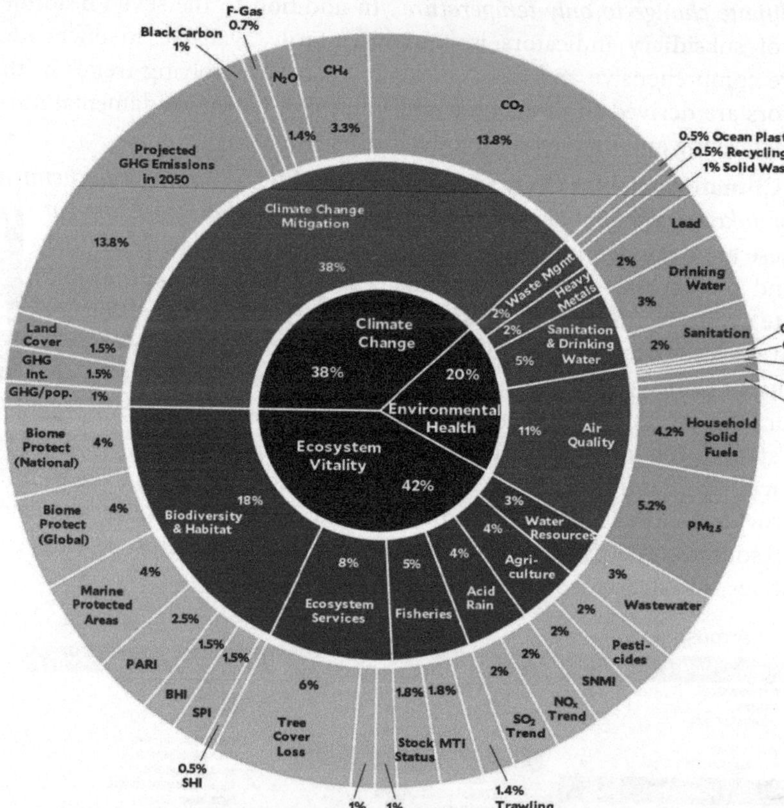

Fig. 4.4: 2022 EPI Framework (source : https://epi.yale.edu)

4.2.3 Environmental Performance Index

The **Environmental Performance Index (EPI)** is a comprehensive assessment that utilizes data-driven analysis to present an overview of sustainability worldwide, by employing 40 performance indicators across 11 issue categories, grouped under three key themes – Climate Change, Environmental Health and Ecosystem Vitality; and ranks 180 countries based on their performance in these three themes. The **40 indicators** serve as a national-scale measure of a country's progress towards established environmental policy targets, and offer a valuable means to identify challenges, establish goals, monitor trends, comprehend outcomes, and recognize effective policy practices. By utilizing accurate data and evidence-based analysis, government officials can refine their policy agendas, engage in meaningful dialogues with stakeholders, and maximize the impact of environmental investments. The EPI contributes to the pursuit of the SDGs as a policy tool and showcases successful initiatives and practices that countries can adopt from their top-performing counterparts.

The EPI ***scores as a whole*** to help to highlight countries excelling in sustainability and draw attention to those lagging, while the detailed disaggregated data that accompanies it provides a more nuanced tool for pinpointing policy deficiencies and anomalies. It is correlated with country wealth, although some countries outperform their economic peers while others lag. However, there are several criticisms to this Index as it relies on a limited set of indicators and metrics to assess a country's environmental performance. This tends to oversimplify complex environmental issues and fails to capture the full scope of a country's sustainability efforts, particularly at the interplay between social, economic, and environmental dimensions of sustainability.

4.2.4 Other Sustainability Measures

There are other foci to sustainability measurement beyond environment and some of these are;

Social Impact Assessment: Evaluates the social and cultural effects of a project or initiative, including factors like employment, community well-being, human rights, and stakeholder engagement.

Economic Indicators: Assess the economic sustainability of a system or project, including metrics such as cost-benefit analysis, return on investment (ROI), and economic value added (EVA).

Corporate Social Responsibility (CSR) Reporting: Measures and reports on an organization's social, environmental, and economic performance, often using globally recognized frameworks like the Global Reporting Initiative (GRI) standards.

Sustainable Procurement: Evaluates the sustainability performance of suppliers and ensures that products and services are procured from environmentally and socially responsible sources.

4.3 INNOVATIONS AND METHODOLOGIES FOR SUSTAINABILITY

Sound methodologies, and decision-making or implementation tools, such as, EIA , SEA, LCA, are of great importance for ensuring overall environmental sustainability.

EIA (Environmental Impact Assessment), Life cycle Assessment (LCA) and SEA (Strategic Environmental Assessment) are **systematic approaches and tools** used to gather and evaluate environmental information before making decisions in the development process. These approaches involve making predictions about how the environment will be affected by various alternative actions and providing guidance on managing those changes if a particular alternative is chosen and implemented. EIA primarily focuses on evaluating the potential environmental impacts of specific physical developments like highways, power stations, water resource projects, and large-scale industrial facilities. LCA provides a framework for measuring the environmental impact of a product across its life and has several tools such as, GaBi, openLCA, SimaPro and Ecochain Mobius. On the other hand, SEA concentrates on assessing proposed actions at a broader level, such as new or modified laws, policies, programs, and plans. It is common for physical developments and projects to be outcomes of the implementation of policies or plans. Thus, an *integrated approach is recommended* (UNEP).

4.3.1 Environmental Impact Assessment (EIA)

The goal of an EIA is to identify and assess the potential positive and negative impacts on the environment, as well as the social and economic aspects associated with the project. The EIA Methodology is as follows:

1. **Screening**: The project plan is screened for scale of investment, location and type of development and if the project needs statutory clearance.
2. **Scoping**: Then the EIA process begins with scoping, where the project's objectives, potential impacts, zone of impacts or boundaries, mitigation possibilities and need for monitoring are identified. This step involves consultations with stakeholders, including government agencies, local communities, and experts, to determine the key issues that should be considered during the assessment.
3. **Baseline Study and Data collection**: A thorough analysis of the existing environmental conditions and socio-economic aspects is conducted, to collect data on air and water quality, biodiversity, land use, cultural heritage, and other relevant parameters, at baseline. The baseline study establishes a benchmark against which potential impacts will be compared.
4. **Impact Prediction (Identification and Assessment)**: Potential environmental impacts associated with the project are identified based on the project's characteristics, such as location, size, and technology used. This step involves analysing the project's activities, processes, and interactions with the environment to determine the potential effects. The identified impacts–Positive and negative, reversible and irreversible, and temporary and permanent impacts–are assessed in terms of their magnitude, duration, and significance. This includes evaluating both the direct and indirect impacts on various environmental components, such as air, water, soil, biodiversity, and human health. Mitigation measures and alternatives may also be considered at this stage.
5. **Mitigation measures and EIA Report:** Based on the impact assessment, potential mitigation measures are identified to avoid, minimize, or compensate for adverse environmental effects. Alternative options, including project design modifications or alternative locations, are also explored to mitigate potential impacts. An Environmental Management Plan (EMP) is developed and reported to guide the implementation, monitoring, and management of the project, which outlines the specific measures and actions to be taken to minimize or mitigate environmental impacts, or the level of compensation for probable environmental damage or loss in adverse situations, throughout the project's lifecycle.
6. **Public Hearing and Consultation**: Throughout the EIA process, public participation and consultation are crucial. Stakeholders, including local communities, NGOs, and interested parties, are given the opportunity to provide input, express concerns, and raise questions related to the project and its potential impacts.
7. **Decision-Making and Approval**: The final EIA report, which includes the findings of the assessment, proposed mitigation measures, and alternatives, is submitted to the relevant regulatory authorities, such as, the Impact Assessment Authority along with the experts, consultants and the project-in-charge, for decision-making. The decision may involve approving the project with conditions, rejecting it, or requesting further information or modifications.
8. **Monitoring and Implementation of EMP**: Once the project is approved and implemented, ongoing monitoring and evaluation are conducted to ensure compliance with the proposed mitigation measures and environmental commitments outlined in the EMP. This step helps

to assess the effectiveness of the EIA process and enables adaptive management to address any unforeseen impacts.

9. **Assessment of Alternatives, Delineation of Mitigation Measures and EIA Report**: Each project should consider and compare various potential alternatives in terms of their environmental characteristics, which should encompass both the location of the project and the technologies employed in its processes. After a thorough review of the alternatives, a mitigation plan should be developed for the chosen option. This plan should be accompanied by an Environmental Management Plan (EMP) that provides guidance to the project proponent on implementing environmental enhancements.
10. **Risk assessment**: The EIA process also includes conducting an inventory analysis and assessing the hazard probability and index as part of EIA procedures.
11. The EIA process may vary in detail and requirements depending on the country, jurisdiction, and specific project characteristics, however, the core objective remains the same.

4.3.2 Life Cycle Assessment (LCA)

LCA is a methodology used to evaluate the environmental impacts associated with a product, process, or system throughout its entire life cycle. There are two different approaches to conducting LCA–the ***'SETAC/EPA Framework for Life Cycle Assessment'***, jointly developed by the

U.S. Environmental Protection Agency (EPA) and the Society of Environmental Toxicology and Chemistry (SETAC); and **ISO 14040**, which is an international standard developed by the International Organization for Standardization (ISO). The choice between the two may depend on factors such as regional applicability, regulatory requirements, and the specific needs of the LCA study.

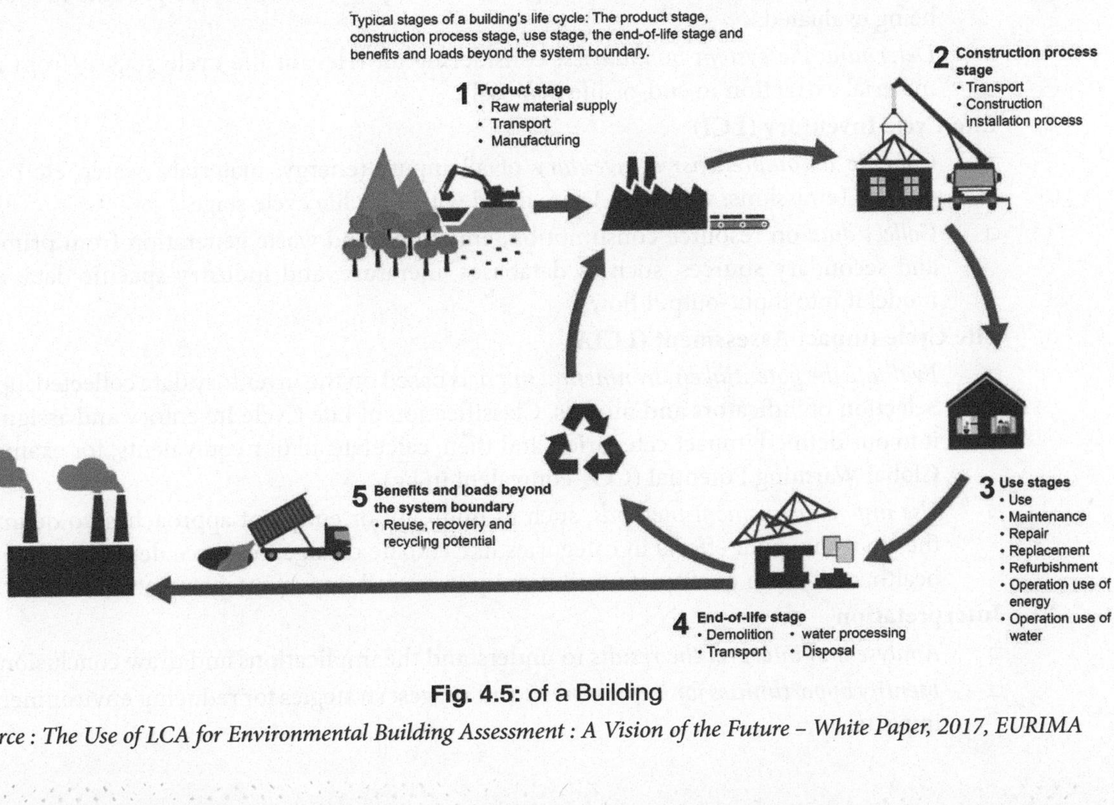

Fig. 4.5: of a Building

source : The Use of LCA for Environmental Building Assessment : A Vision of the Future – White Paper, 2017, EURIMA

While they share common principles and goals, there are few differences, such as;

- ***Coverage and Scope:***
  - The SETAC/EPA framework emphasizes the need to consider the full life cycle of a product or system, from cradle to grave, including raw material extraction, manufacturing, use, and end-of-life.
  - ISO 14040 also advocates for a comprehensive life cycle perspective but allows flexibility in defining the scope of the study based on the intended application and goals.
- ***Methodology:***
  - The SETAC/EPA framework provides more specific guidance on conducting impact assessment, including midpoint and endpoint approaches.
  - ISO 14040 provides general principles and guidelines but does not prescribe specific impact assessment methodologies, allowing flexibility in selecting appropriate methods.
- ***Application:***
  - The SETAC/EPA framework is commonly used in North America and may be more aligned with regulatory requirements in that region.
  - ISO 14040 is an international standard and is widely adopted globally, accommodating different regional contexts and requirements.

The ISO 14040 outlines a **four-step process** for conducting an LCA, as elaborated below;

1. **Goal and Scope Definition**
   - *Define the purpose and boundaries of the LCA study*, including the product or system to be assessed and the specific environmental impacts of interest.
   - *Identify the functional unit*, which quantifies the performance of the product or system being evaluated.
   - *Determine the system boundaries*, considering all relevant life cycle stages (from raw material extraction to end-of-life disposal).
2. **Life Cycle Inventory (LCI)**
   - *Compile a comprehensive inventory* of all inputs (energy, materials, water, etc.) and outputs (emissions, waste, etc.) associated with each life cycle stage.
   - *Collect data* on resource consumption, emissions, and waste generation from primary and secondary sources, such as databases, literature, and industry-specific data, and model it into input-output flows.
3. **Life Cycle Impact Assessment (LCIA)**
   - *Evaluate the potential environmental impacts* based on the inventory data collected, upon Selection of indicators and models, Classification of Life Cycle Inventory and assigning it to our defined impact categories, and then, calculate all our equivalents, for example, Global Warming Potential (CO_2-equivalent in kg).
   - *Use impact assessment methods*, such as midpoint or endpoint approaches, to quantify the environmental effects in categories like climate change, resource depletion, human health, ecosystem quality, etc. and sum up in overall impact category totals.
4. **Interpretation**
   - *Analyse and interpret the results* to understand the implications and draw conclusions.
   - *Identify opportunities for improvement* and suggest strategies for reducing environmental impacts.

- *Consider the limitations and uncertainties* of the study and communicate the findings accurately and transparently.

Additional considerations;

- ***Sensitivity analysis*** to assess the influence of different parameters or assumptions on the results.
- ***Reporting*** through a comprehensive report summarizing the study methodology, results, and conclusions, and ***Communicating the findings*** to stakeholders and facilitate informed decision-making.

4.3.3 Strategic Environmental Assessment (SEA)

SEA helps incorporate environmental considerations into strategic planning and policy-making processes, and ensures that environmental and social impacts are systematically assessed and integrated into decision-making, promoting sustainable development, through the following steps;

1. **Initiation and Scoping**
   - *Determine the need for SEA* and establish the objectives and scope of the assessment.
   - *Identify the key decision-makers and stakeholders* who should be involved in the SEA process.
   - *Define the legal and institutional framework* for conducting the SEA.
2. **Baseline Assessment**
   - *Collect and analyse information* about the existing environmental, social, and economic conditions.
   - *Identify the potential environmental effects* associated with the plan, policy, program, or strategy under consideration.
   - *Consider relevant environmental policies*, legislation, and international commitments.
3. **Setting Objectives and Developing Alternatives**
   - *Establish environmental objectives and targets* that align with sustainable development goals.
   - *Generate a range of alternative options* or scenarios that could achieve the objectives.
   - *Assess the potential environmental and social impacts* associated with each alternative.
4. **Impact Assessment**
   - *Evaluate the potential effects* of each alternative on the environment, including direct and indirect impacts.
   - *Analyse the cumulative effects* of multiple projects or activities when appropriate.
   - *Identify the key environmental and social issues* that need to be addressed during implementation.
5. **Mitigation and Enhancement Measures**
   - *Develop measures to avoid, minimize, or mitigate* adverse environmental and social impacts.
   - *Explore opportunities* to enhance positive environmental and social outcomes.
   - *Consider alternative approaches*, technologies, or management strategies that promote sustainable development.

6. **Integration and Decision-Making**
   - *Integrate the findings* of the SEA into the decision-making process for the plan, policy, program, or strategy.
   - *Communicate the results* of the assessment to decision-makers, stakeholders, and the public.
   - *Consider the SEA recommendations* and findings alongside other relevant factors in the decision-making process.

7. **Monitoring, Review, and Adaptation**
   - ***Establish a monitoring and review mechanism*** to track the implementation of the plan, policy, program, or strategy.
   - ***Evaluate the effectiveness of the SEA*** recommendations and measures over time.
   - ***Incorporate adaptive management practices*** to ensure continuous improvement and responsiveness to changing circumstances.

Overall, these methodologies and tools are critical in ensuring that environmental considerations are integrated into decision-making processes at different levels, from strategic planning to project implementation, by promoting sustainable development through stakeholder engagement.

UNIT SUMMARY

This unit on environmental (impact) discusses the nuances between the observable phenomena of global warming and its potential connection to overall climate change. Firstly, it elucidates the foreboding impact of pollution and the various strategies and policies for mitigation. Secondly, it explains the challenges and uncertainties in monitoring and predicting such changes due to the non-stationary nature of the variables. It further elaborates on environmental monitoring, metrics, indicators, etc to better define the state of the environment and its performance. Finally, the unit explicates the various methodologies and tools innovated to quantify environmental impact, in order to achieve sustainable development goals and instil urgency and understanding in stakeholders for immediate action.

EXERCISES

I. Multiple Choice Questions

Q.1 Which of the following are greenhouse gases (GHG)?

a. all of the below
b. Methane (CH4)
c. Nitrous oxide (N2O)
d. Carbon di oxide (CO2)

Q.2 What is the percentage of CO2 in the global GHG emissions?

a. more than 60%
b. 76-78%
c. less than 40%
d. 50%

Q.3 Which of the following sectors is the fastest growing source of GHG emissions (since 1990) ?

a. Industrial processes
b. Transportation
c. Electricity and heating
d. Agriculture and Land Use

Q.4 What is the most important indicator with respect to climate change?
   a. Surface Temperature and Ocean Heat (Temperature and Energy)
   b. Ocean Acidification and Sea-level
   c. Glaciers and Arctic and Antarctic Sea Ice Extent (Cryosphere)
   d. all of the above

Q.5 Which of the following are referred as 'true sustainability metrics'?
   a. Economic, ecological, and sociological indicators
   b. Sustainability indicators
   c. Socio-economic, eco-efficiency, and socio-ecological indicators
   d. Environmental Performance Index

Answers of Multiple Choice Questions				
4.1 **a**	4.2 **b**	4.3 **a**	4.4 **d**	4.5 **b**

II. Short and Long Answer Type Questions

Q.6 What is Global Warming? Briefly explain the phenomena and its impact on environment.

Q.7 What are some of the potential GHG emissions and pollution mitigation measures presently being employed?

Q.8 Discuss the various types of environmental monitoring and tools/methods used for each.

Q.9 What are the various types of 'metrics and indicators' for denoting the state of the environment? How are these helpful in predicting environmental performance and impact, elucidate with examples.

Q.10 What are some of the innovative methodologies and tools for assessing environmental impact? Illustrate anyone that will be most useful to the profession of civil engineering.

KNOW MORE

About World Greenhouse Gas emission & World Emissions Clock

https://www.wri.org/insights/4-charts-explain-greenhouse-gas-emissions-countries-and-sectors

https://www.epa.gov/ghgemissions/sources-greenhouse-gas-emissions

https://www.iea.org/reports/co2-emissions-in-2022#

Climate.nasa.gov

https://www.brookings.edu/blog/future-development/2022/11/29/tracking-emissions-by-country-and-sector/

About Sustainable agriculture

https://wedocs.unep.org/bitstream/handle/20.500.11822/25950/sustainable_agriculture.pdf?sequence=1&isAllowed=y

About Circular Economy

https://buildingcircularity.org

The Paris Agreement

https://unfccc.int/process-and-meetings/the-paris-agreement

International Environmental Law

https://wedocs.unep.org/bitstream/handle/20.500.11822/21491/MEA-handbook-Vietnam pdf?sequence=1&isAllowed=y

REFERENCES AND SUGGESTED READINGS

Books and Publications

- [Book] Fischer, T. B. (2010). *The theory and practice of strategic environmental assessment: towards a more systematic approach*. Routledge
- [Book] Glasson, J., & Therivel, R. (2013). *Introduction to environmental impact assessment*. **Routledge.**
- [UN Report] Sachs, J., Kroll, C., Lafortune, G., Fuller, G., & Woelm, F. (2022). *Sustainable development report 2022*. Cambridge University Press
- Bueno de Mesquita, C. P., White, C. T., Farrer, E. C., Hallett, L. M., & Suding, K. N. (2021). *Taking climate change into account: Non-stationarity in climate drivers of ecological response. Journal of Ecology*, *109*(3), 1491-1500.
- Sikdar, S. K. (2003). *Sustainable development and sustainability metrics. AIChE journal*, *49*(8), **1928-1932**
- UNEP (2012) *Trade and Environment Briefings: Sustainable Agriculture*, **2012**
- UN (2017) *Report on the Rule of Law and Transitional Justice in Conflict and Post-Conflict Societies*

Web sources

- https://www.eea.europa.eu/policy-documents/com-2015-0614-final
- https://www.causeweb.org/usproc/sites/default/files/usclap/20181/Statistical_Analysis_of_Climate_Change_Variables.pdf
- https://unfccc.int/process-and-meetings/the-paris-agreement
- https://wedocs.unep.org/bitstream/handle/20.500.11822/25950/sustainable_agriculture.pdf?seque nce=1&isAllowed=y
- https://public.wmo.int/en/programmes/global-climate-observing-system/essential-climate-variables

05 BUILT ENVIRONMENT

UNIT SPECIFICS

Through this unit we have discussed the following aspects:

- ***The Built Environment and Its Impact***
  - *Facilities management & Sustainable Facility management strategies*
  - *Building Control systems–Climate control systems, Security systems*
  - *Energy efficient built environments & Recycling*
  - *LEED Rating & other Indian rating systems*
  - *Intelligent/ Smart Buildings*
- ***Aesthetics of built environment***
  - *Role of Urban Arts Commissions*
  - *Heritage Conservation, Structural Repair and Rehabilitation, Retrofitting*
- ***Innovations and methodologies for ensuring Sustainability***

Besides giving a large number of multiple choice questions as well as questions of short and long answer types marked in two categories following lower and higher order of Bloom's taxonomy, a list of references and suggested readings are given in the unit so that one can go through them for practice.

There is a "Know More" section, which has been carefully designed so that the supplementary information provided in this part becomes beneficial for the users of the book. It is important to note that for getting more information on various topics of interest some QR codes have been provided which can be scanned for relevant supportive knowledge This section mainly highlights applications of the subject matter for our day-to-day real life or/and industrial applications on variety of aspects, case study related to environmental, sustainability, social and ethical issues whichever applicable, and finally inquisitiveness and curiosity topics of the unit.

RATIONALE

This elementary unit on the built environment and its role to mitigate environmental impact, discusses the various practicalities, such as, Facility management, Building control systems, Building Rating, and Codes and Standards, to impart state-of-the-art knowledge to improve the civil engineers' technical know-how, and perhaps open up possibilities of pursuing future expertise in one of the several areas discussed.

UNIT OUTCOMES

List of outcomes of this unit is as follows:

U5-O1: Knowledge on the various aspects of Built Environment and its impact

U5-O2: Understanding of Aesthetics and its importance for cultural heritage

U5-O3: Knowledge on innovations and methodologies for ensuring environmental Sustainability (Codes and Standards)

Unit-5 Outcomes	EXPECTED MAPPING WITH COURSE OUTCOMES (1- Weak Correlation; 2- Medium correlation; 3- Strong Correlation)						
	CO-1	CO-2	CO-3	CO-4	CO-5	CO-6	CO-7
U5-O1	3	3	2	2	3	3	2
U5-O2	2	2	3	2	2	3	3
U5-O3	3	3	3	2	3	3	3

Our engineered, built-environment sits nestled in the natural environment, which is presently undergoing several climatic changes leading to rising concerns and potential risks to human life. The built-environment is defined by the U.S. Environmental Protection Agency (EPA, 2021) as *"human-made or modified surroundings in which people live, work, and recreate"*, to which the National Academies (2017) adds, *"surroundings that provide the setting for human activity, ranging in scale from buildings and parks or green space to neighbourhoods and cities that can often include their supporting infrastructure, such as water supply or energy networks"*.

5.1 THE BUILT ENVIRONMENT AND ITS IMPACT

The major burden of societal and global impact is created and experienced at the intersection of "*buildings, cities, and urban spaces, and the ways people use them*" (WHO, 2016), i.e., intersection of 'Spaces and Infrastructure', comprising of design, construction, occupancy management, building operations, maintenance, furniture and equipment, etc.; and 'People and Organisation', consisting of health services, catering, event management, hospitality, security, safety, human resources, logistics, document management, and accounting. The concept of "**sustainable construction**" integrates environmental criteria into the entire lifecycle of a project, including building, maintenance, and eventual demolition, suggests various factors to be emphasized during the construction phase, beyond material choices, such as techniques to minimize dust, noise, and soil and water contamination, waste management measures throughout the construction, use, and demolition phases, and efforts to enhance material reuse and recycling.

Buildings alone contribute to approximately 40% of energy and process-related CO2 emissions, 50% of all extracted materials, 33% of water consumption, and 35% of generated waste, and additionally, contribute towards resource depletion, pollution of air, water, and land, as well as loss of biodiversity. Of which, the use/operational phase leads to the highest environmental impact across the entire life of the building and is known to be caused by the users and the physical characteristics of the building, i.e., energy and water use and waste generation. Therefore, it is important to employ *systems thinking and the lifecycle view* to address impending issues of environmental impact mitigation through incorporating sustainability strategies, beyond design and construction, into facility management.

5.1.1 Facilities Management

Facilities management is a vital aspect of managing construction businesses, as it involves strategic planning and thoughtful consideration of daily operations. British architect and Editor of Architectural Association Journal (1965-67) and Facilities (1982-90), Francis Duffy explored the relationship between organisational structure and office layouts in his doctoral work during the 1960s, and later, pioneered 'Space Planning and Facility management' in Europe. At the time, the term was propounded by Ross Perot, founder of Electronic Data Systems, USA, when the introduction of computers and IT infrastructure demanded design changes in the 1970s.

Facility management entails the effective management of physical workspaces–buildings, infrastructure, and other physical assets, to support the delivery of services and achieve organizational objectives. Civil engineers are ideal for this role due to their technical skills and good communication, and develop leadership in the upcoming area of facility management. Facility Management integrates the principles of business administration, architecture, and the behavioural and engineering sciences; and involves a wide range of activities, including maintenance and operations, space planning, asset management, health and safety, energy management, and sustainability.

The **key activities** involved are:

- Asset Management
- Master Planning
- Space Management
- Estates Strategies
- Maintenance, cleaning, testing, and inspection
- Refurbishment, retrofitting, and renovation
- Acquisitions and sales
- Procurement and project management
- MEP (Mechanical, Electrical and Plumbing) and Technical services
- Contract Management
- Sustainability
- Budget Management
- Brand Management
- Quality Assessment
- Regulatory Compliance
- Asset exploitation and income generation
- Ensuring the continuity of the business
- Delivery of new technologies
- Bringing change and efficiency in work practices
- Safety and security
- Traffic, transport, and parking

Role of Facility manager responsibility towards sustainability

In the field of facilities management, the focus has shifted to the impacts of climate change, encompassing not only the environmental effects of buildings but also their resilience and capacity to address associated issues. The primary challenge lies in the increasingly severe weather conditions and unpredictable shifts in weather patterns, such as tides, cold spells, floods, storms and droughts, and prolonged seasons, making it necessary for facilities managers to transition from preventive measures to predictive strategies, like enhanced service level requirements and disaster recovery plans for ensuring business continuity. Facility managers have **three significant responsibilities** in addressing climate change:

- ***Conduct assessments to evaluate the vulnerability of current facilities*** and their ability to adapt to climate change. This involves creating plans and procedures to address climate-related challenges and ensure effective responses.
- ***Assist in the development of new structures and buildings*** that align with the changing environmental landscape. This includes reviewing designs and incorporating sustainable building materials to promote environmentally friendly practices.
- ***Develop a forward-looking strategy to manage the evolving dynamics of the built environment*** due to shifts in climate change regulations and compliance requirements. This involves anticipating future changes and implementing measures to ensure ongoing environmental sustainability.

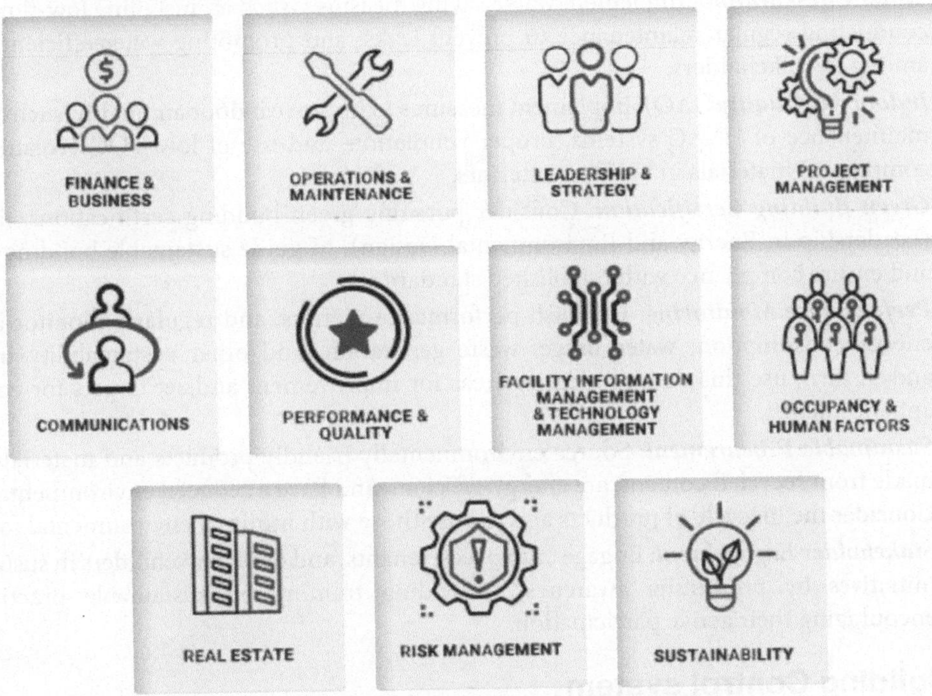

Fig. 5.1: 11 Core competencies of Facility Management

source : www.ifma.org

Sustainability Facility management strategies

Sustainable facility management (SFM) is "*a unique process that offers a facility manager the authority to make structural, architectural, and operational changes to reduce the negative impact of buildings on their occupants and the environment*" (Fennimore, 2014). Implementing sustainable practices in facility management to mitigate climate change risks requires a comprehensive approach that considers various aspects of operations. Adopting a lifecycle approach in decision-making processes, taking into account the environmental impacts of products and systems from extraction to disposal, and considering factors such as, durability, recyclability, and end-of-life management are essential towards enabling sustainable facility management.

The following are some **practical strategies** for implementing sustainable practices in facility management:

1. ***Energy Efficiency***: Implement energy-saving measures, such as, installing energy-efficient lighting, utilizing programmable thermostats, optimizing HVAC systems, and conducting energy audits to identify areas for improvement.
2. ***Renewable Energy Integration***: Explore opportunities to incorporate renewable energy sources, such as solar panels or wind turbines to reduce reliance on fossil fuels and decrease carbon emissions.
3. ***Waste Management***: Develop a waste management plan that includes recycling programs, waste reduction initiatives, proper disposal methods, and encourage all stakeholders to reduce, reuse, and recycle materials.
4. ***Water Conservation***: Implement water-saving measures, such as, installing low-flow fixtures, conducting regular maintenance to prevent leaks, and promoting water-efficient practices among all stakeholders.
5. ***Indoor Air Quality*** (IAQ): Implement measures to improve indoor air quality, such as, regular maintenance of HVAC systems, proper ventilation, and use of low VOC (volatile organic compound) materials in building interiors.
6. ***Green Building Certification***: Consider pursuing green building certifications, like LEED (Leadership in Energy and Environmental Design), to guide sustainable building practices and ensure compliance with established standards.
7. ***Performance Monitoring***: Establish performance metrics, and regularly monitor and track energy consumption, water usage, waste generation, and other sustainability indicators, and in turn, use this data to identify areas for improvement and set targets for continuous enhancement.
8. ***Sustainable Procurement***: Source environmentally friendly products and materials that are made from recycled content, are energy-efficient, and have a reduced environmental impact. Consider the lifecycle of products and choose those with minimal environmental footprints.
9. ***Stakeholder Engagement***: Engage employees, tenants, and other stakeholders in sustainability initiatives by promoting awareness, providing training on sustainable practices, and encouraging their active participation.

5.1.2 Building Control system

A building control system is a centralized system that monitors and controls various building systems and equipment to ensure efficient operation, occupant comfort, and safety. These systems are commonly referred as; Energy Management Systems (EMS), which is responsible for managing environmental functions within a building; Building Automation Systems (BAS), which focuses on controlling technical automation processes; and Building Management Systems (BMS), which encompasses a broader range of functions, including providing status reports on environmental conditions, monitoring elevator operations, and tracking the location of individuals for security purposes.

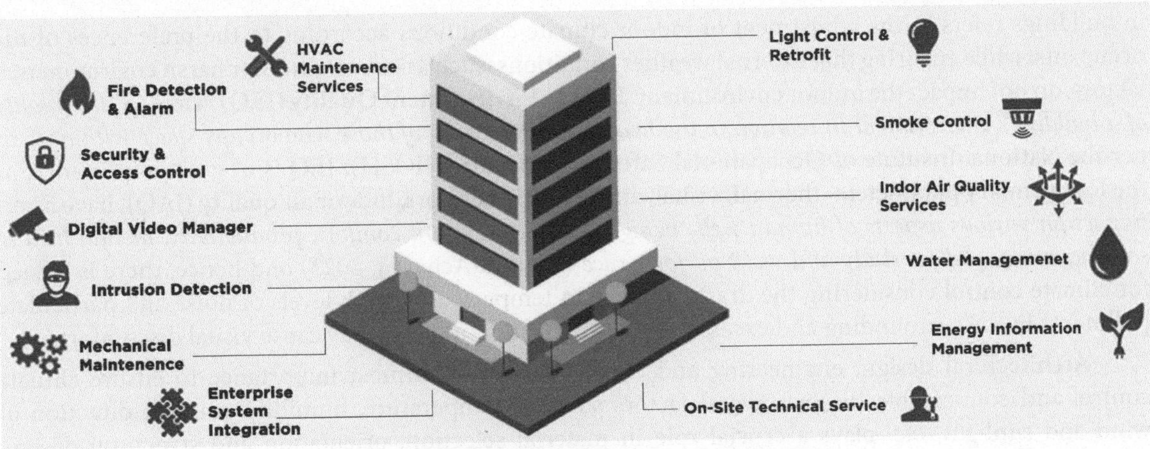

Fig. 5.2: Elements of a Building Control System

source : www.truviconline.com)

The major functions are, as follows;

- ***Data Analysis and Reporting:*** Building control systems provide data analysis and reporting capabilities, enabling facility managers to assess building performance, identify trends, and make informed decisions for improving energy efficiency and operational effectiveness.
- ***Monitoring and Control:*** Monitoring systems collect data on various parameters such as temperature, humidity, occupancy, and energy consumption to monitor the building's performance and ensure it operates within predefined limits. Based on the collected data, the system can automatically adjust and control the building systems to maintain optimal conditions. For example, it can regulate HVAC systems to provide comfortable temperatures or adjust lighting levels based on occupancy.
- ***Management and Decision-aid:*** Building control systems help optimize energy usage by tracking energy consumption, identifying inefficiencies, and implementing strategies for energy conservation, such as scheduling equipment operation and implementing demand response programs. The system also generates alerts and notifications in case of equipment malfunctions, abnormal conditions, or security breaches, allowing facility managers to take prompt action.

EMS, BAS, and BMS consists of a network of sensors, controllers, and software that communicate and coordinate the functioning of different building systems. EMS collects data from sensors throughout the building and uses the information to make adjustments and optimize the performance of these systems. While **BMS/BAS typically integrates and controls systems** such as heating, ventilation, and air conditioning (HVAC), lighting, access control, fire alarms, security systems, and energy management. The key difference is, while an **EMS is primarily a monitoring tool** at a micro-level, gathering detailed device-level telemetry data from the deployed sensors and analysing it, the BMS/BAS, often used synonymously, works at a macro-**level by providing high-level, real-time control, such as, regulate the temperature, humidity or pressure, or prevent smoke from spreading throughout a building.**

Building Climate Control System

A building and its indoor spaces are constantly influenced by both internal and external factors that can disrupt the indoor climate, such as heat, air emissions, and changes in humidity levels. Climate control in buildings refers to the adjustment of indoor climate conditions according to the preferences of the occupants, while ensuring that external weather conditions, such as extreme heat or harsh environmental factors, do not impact the indoor environment. **Indoor Environment Quality** (IEQ) refers to "*the quality of a building's environment in relation to the health and wellbeing of those who occupy space within it*", as per the National Institute of Occupational Safety and Health (NIOSH). IEQ is of great importance, as the four primary parameters–thermal, visual, and acoustic comfort, indoor air quality (IAQ), have *direct impact on various aspects of human well- being, such as, occupant comfort, productivity, mental health, satisfaction, as well as study and work performance* (Ali and Acharya, 2023) and hence, there is a need for climate control considering the drastic change in temperatures, high levels of noise and particulate pollutants in the surrounding and excess use of artificial lighting that may cause visual discomfort.

Architectural design, engineering and construction is of utmost importance to ensure climate control and comfortable IEQ as several factors such as, temperature, humidity, rainfall, direction of wind and sunlight, etc. plays a crucial role in material selection, orientation and structural design. Air temperature affects the exterior of the building while the soil temperature affects the foundation. Humidity is a very important consideration for the structural integrity of the materials, and anti-corrosive and water-resistant materials are preferred in areas with higher levels of humidity. In areas with heavy rainfall, sloping roofs are preferred as opposed to flat roofs, so that the integrity of the structure is not compromised. Sunlight is also an important consideration, as the wall to window ratio increases or decreases according to need of the climate. Wind load is also considered while designing, since it has the capability to damage the building structurally. **Active and Passive design strategies** are employed to improve IEQ and liveability of a building. '*Active*' implies the use of mechanical and electrical systems, such as, air-conditioning, heat pumps, radiant heating, heat recovery ventilators, and electric lighting which uses purchased energy, while '*Passive*' refers to harnessing available ambient energy through techniques such as, maximizing natural light (daylighting), utilizing natural airflow for ventilation, and garnering solar energy. Some **passive design strategies** are;

- ***Seeking inspiration from Vernacular architecture***. It is the ancient traditional architectural design and construction method of a civilisation, and usually it is incorporated with unique elements and techniques that controls the natural climate of that region and suitable of the life of the indigenous people. Introduction of verandahs and courtyards are one such design element borrowed from vernacular architecture that has the ability to reduce heat and allow cross-ventilation.
- ***Incorporating Bioclimatic design***, which aims to maximize the utilization of local climate data and leverages local environmental resources like sunlight, wind, air, vegetation, soil, and sky to minimize energy consumption. Proper thermal insulation of the external envelope of the building and air-tight structures; use of solar energy for heating in winter time, achieved by a proper passive solar heating system; removal of the heat from the building by passive cooling systems and natural ventilation; Protection of the building from overheating by proper shading solutions and vegetation in summer time; and the use of solar energy for day-lighting all year round, are some of the recommendations by EU. It may be argued that this is also seen in vernacular architecture.
- ***Employing shading devices***, such as, fin walls, vertical or horizontal shading overhang devices, and extended roof overhangs, which are integrated into the building structure, and external shading devices, such as, eaves, awnings, screens and shutters, louvres, verandahs, pergolas, and shading with foliage, trees, and shrubs.

- ***Introducing rigid air barrier, thermal insulation and sound attenuating interior lining*** to brick or concrete masonry or to high-density sheet cladding for exterior walls and building envelope, using concrete or tiled roofing, and affixing sound-attenuating external windows and doors, to reduce external noise and overall, acoustic discomfort.

'*Active*' control systems leverage technologies, such as, ICT, cloud computing, wireless sensors, and tools, such as, BMS and BAS to help improve climate control efficiency.

Building Security System

The **building security system** is specifically developed to oversee and regulate mechanical and electrical installations, fire protection and escape mechanisms, as well as address issues related to burglary, assault, and emergency communication.

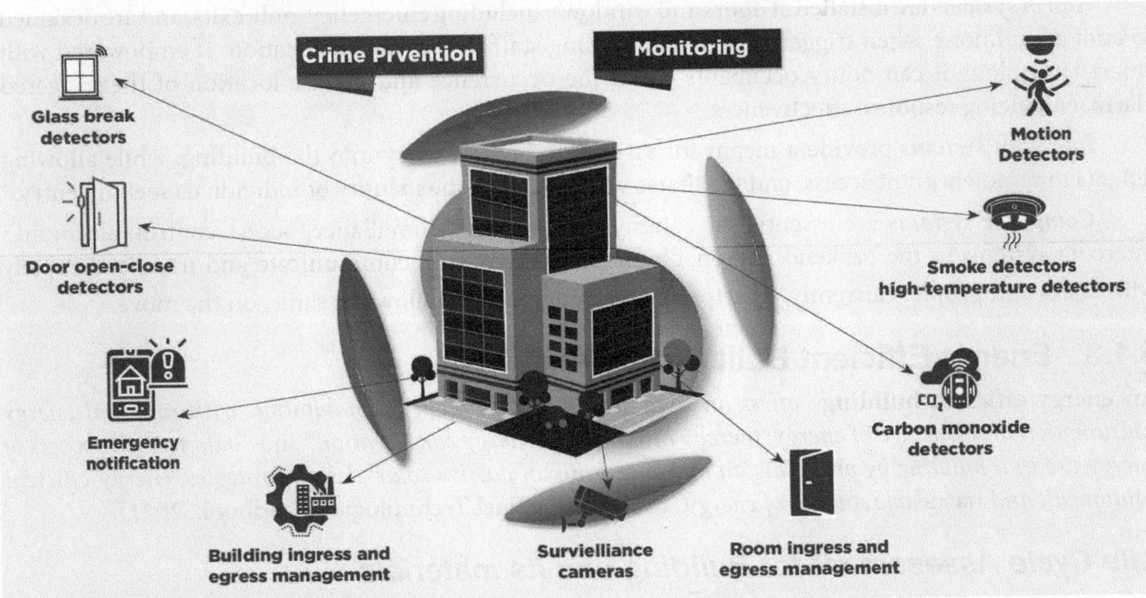

Fig. 5.3: Features of a Building Security System

source:www.truviconline.com

In high-rise buildings and large complexes which are densely occupied, ensuring effective fire safety measures is of utmost importance. To ensure **fire safety** as it is necessary to monitor and manage the following:

- Detection and suppression of fires,
- Ensuring the movement and safety of individuals,
- Controlling smoke through pressurization and barriers,
- Establishing safe places of refuge, and
- Facilitating emergency arrangements and communication.

An **effective security system** has the following components:

Video surveillance systems comprises of the placement of multiple cameras in strategic locations both inside and outside a building, to serve as a deterrent to potential criminals, and to provide valuable evidence for police investigations and insurance claims. Furthermore, modern video surveillance systems can be integrated with other security technologies. For instance, you can have a surveillance camera positioned to monitor a keypad entry system at the front door, which can enable detection of any tampering or misuse of the entry device.

Access control encompasses the procedures and mechanisms that regulate entry and restrict unauthorized access to buildings, rooms, or specific areas, and is a fundamental aspect of maintaining security in any location. To ensure optimal security, it is essential to have a cohesive access control system in place, as it is inconvenient for occupants to use different Access control technologies, such as, separate key fobs, PIN codes, or key cards, to access different areas within the building. Therefore, it is recommended to synchronize the access control system throughout the building, preferably by using a unified system provided by the same security system company.

Alarm systems are installed at doors and windows, including emergency-only exits, and are designed to emit a loud noise when triggered, alerting building staff to the specific location. If empowered with smart technology it can notify occupants about the occurrence and precise location of the triggered alarm, enhancing response effectiveness.

Intercom systems provide a means for visitors to request entry into the building, while allowing tenants to remotely grant access, and facilitates verification of the identity of individuals seeking entry.

Computer systems are essential to cohesively tie up the surveillance, access control, alarm and intercom systems in the backend and enable the user to control, communicate and interact remotely with assets and people. Currently, smartphones are integrated to allow the same, on the move.

5.1.3 Energy Efficient Built Environment

An **energy efficient building** "*offers an appropriate environment for habitation with minimal energy consumption and wastage of energy, thereby maximizing energy conservation*" and "*balances all aspects of energy use in a building by providing an optimized mix of passive solar–design strategies, energy-efficient equipment, and renewable sources of energy*" (Sustainable Fuel Technologies Handbook, 2021).

Life Cycle Assessment for Building and its materials

Life Cycle Assessment (LCA) based on ISO14040 (2006) is being increasingly adopted in architectural design and construction, as it CA enables decision-making processes, such as, evaluating and selecting environmentally-friendly products and optimizing construction processes. In addition to assessing environmental impacts, LCA also facilitates estimating life cycle costs (LCC) and conducting life cycle energy analysis (LCEA). The incorporation of LCA into the building sector supports sustainable decision-making and promotes the evaluation of environmental, economic, and energy-related aspects throughout the life cycle of buildings. It is noteworthy that energy efficiency has been a primary focus in the environmental design of buildings, with the **use/operational phase** playing a dominant role in the LCA of buildings, due to the high energy demand associated with building operation. Operating energy, which includes energy for HVAC, domestic hot water, lighting, appliances, and building maintenance, is easily quantifiable and can account for up to 85% of total energy consumption and 70-90% of the environmental impact. However, many of these factors are beyond the control of designers, particularly during the conceptual stage, as they are spread over the decades-long lifespan of a building.

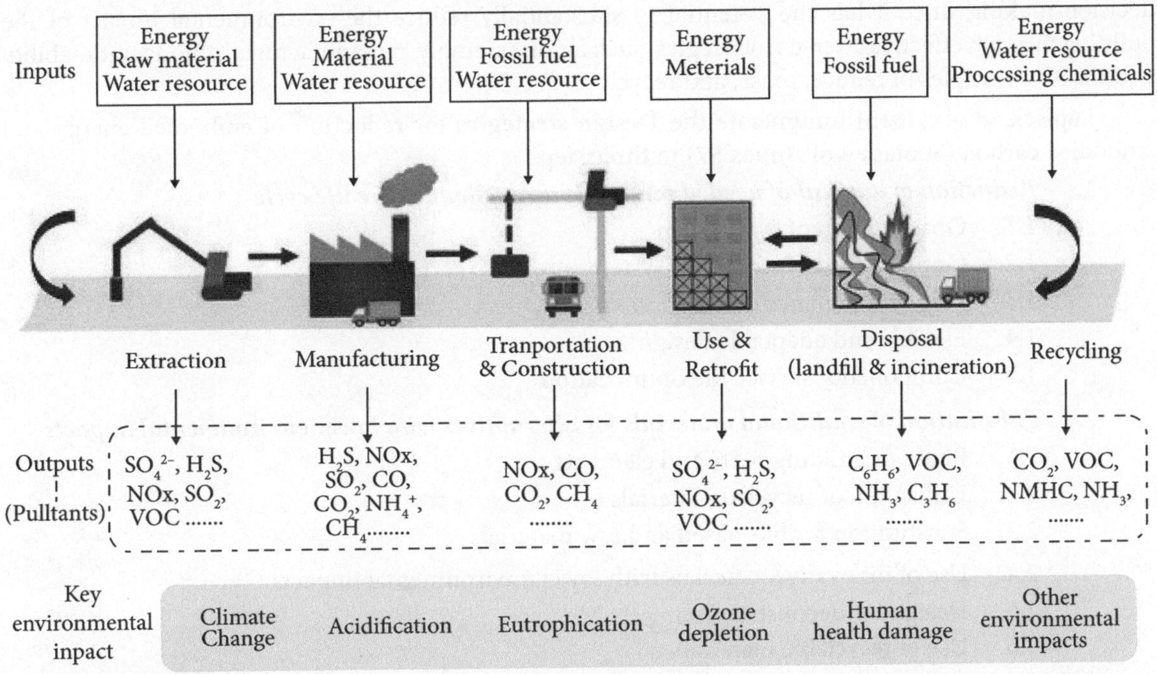

Fig. 5.4: Key Environmental Impacts during Life Cycle of Building Materials (Huang, et al., 2020)

Therefore, the selection of appropriate materials, which is often emphasized as a directive for energy- efficient design, becomes a critical decision factor in the design process. It is recommended to consider material selection as early as possible, as it not only impacts the operational energy performance of the building but also influences the total embodied energy and potential environmental impact.

Embodied Energy and Embodied Carbon

From a life cycle perspective, the environmental impact and energy consumption of buildings are closely linked to the choice of materials. However, there is often a lack of alignment between the service life of materials (which determines the intervals for material renewal) and the service life of buildings (the duration of the operational phase). This can result in a *trade-off between material selection and building energy demand* and so, requires a holistic approach (Heeren et al., 2015) that considers the effects and trade-offs. For example, in a residential building, since the 'wood variant' is susceptible to solar and internal gains, and the 'concrete variant' isn't, the energy spent on demand for space cooling of wood is higher than concrete in spite of wood having lower environmental impact than concrete. However, for cool climates, a wooden structure and this very same property will be beneficial and energy-saving.

Every building material possesses an **'Embodied Energy' or Embodied Carbon,** which refers to the sum impact of all greenhouse gas emissions attributed to that material during its life cycle from extraction, till disposal. The after-use life cycle phase plays a significant role in determining the overall embodied energy of a building. LCA for the "end of life" (EOL) evaluates the environmental impact, including the energy required for demolition, at the end of a building's service life, as this is vital for decision-making since it has the potential to substantially reduce the environmental impact of the building through effective after-use strategies, such as, disassembly, remanufacturing, biodegrade-ability and the 3R principles of reduce, reuse, and recycle .

Lupíšek, et al. (2015) innumerate the **Design strategies** for reduction of embodied energy and embodied carbon (Subtask 4 of Annex 57) in three steps:

1. ***Reduction of amount of needed materials throughout entire life cycle***
   - 1.1. Optimization of layout plan
   - 1.2. Optimization of structural system
   - 1.3. Low-maintenance design
   - 1.4. Flexible and adaptable design
   - 1.5. Components' service life optimization
2. ***Substitution of traditional materials for alternatives with lower environmental impacts***
   - 2.1. Reuse of building parts and elements
   - 2.2. Utilization of recycled materials
   - 2.3. Substitution for bio-based and raw materials
   - 2.4. Use of innovative materials with lower environmental impacts
   - 2.5. Design for deconstruction
   - 2.6. Use of recyclable materials
3. ***Reduction of construction stage impact***

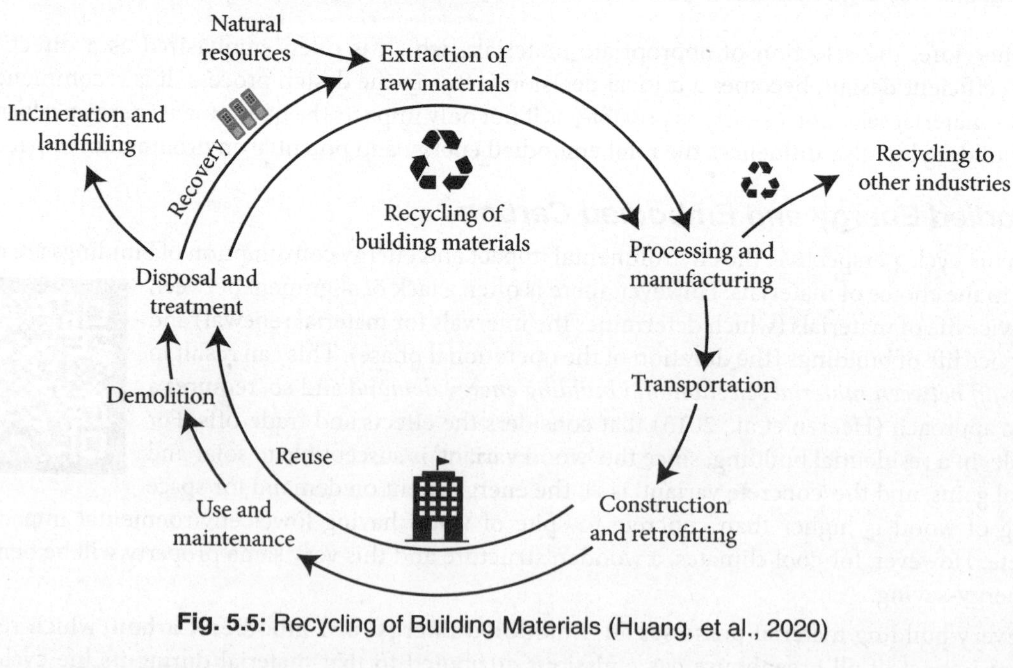

Fig. 5.5: Recycling of Building Materials (Huang, et al., 2020)

Building Materials and Recycling

A large number of building materials are **reusable and recyclable** (Kralj & MariČ, 2008) , such as, Wood and untreated timber, and Earthen Materials (*reusable/recyclable/biodegradable*); Metals, mainly Steel,

Aluminium, Iron, Copper, Masonry and Bricks, and Glass and ceramic (*reusable/recyclable*); Asphalt and Concrete (*may be crushed and recycled*); and Gypsum/Drywall (*recyclable, sometimes biodegradable*). Some innovative, modern recycled building materials are ; **Plant-Based Polyurethane Foam Boards**, possessing good heat transfer resistance, quality sound insulation, and mould and insect resistance; **Hempcrete**, a sustainable, carbon negative alternative to concrete bound with hemp fibres; **Ecobrick**, a plastic bottle filled with small plastic waste; **Rammed Earth**, made of compacted, excavated earth from the construction site, mixed with concrete; **Chip[s]** board, is a fibre-reinforced bioplastic made from potato waste.

Recycling materials can contribute greatly towards **reducing impact** and can be achieved by;

- ***Identifying materials in existing buildings*** that can be salvaged and reused in new construction projects may include, structural elements, fixtures, flooring, doors, and windows.
- ***Establishing on-site recycling programs*** to process and reuse construction and demolition waste, and setting up in dedicated areas for sorting and separating different materials such as concrete, metal, wood, and plastics.
- ***Implementing comprehensive waste management plans*** for construction sites, with clear guidelines for sorting, separating, and recycling different types of waste generated during the construction process.
- ***Participating in material exchange networks*** or online platforms where builders, contractors, and suppliers can connect to exchange surplus materials.
- ***Collaborating with local recycling centers***, waste management facilities, and other stakeholders to establish efficient recycling systems for construction materials.

Incorporating design principles of **'Design for Disassembly'** make it easier to disassemble and separate materials during the end-of-life phase, by using modular construction techniques and joinery systems, for easy dismantling without damaging the materials, is a potential approach to carefully deconstruct buildings to preserve reusable materials.

5.1.4 LEED Rating

LEED, which stand for Leadership in Energy and Environmental Design, is a green building rating system developed by the US Green Building Council (USGBC). In India, Confederation of Indian Industry (CII) formed the Indian Green Building Council (IGBC) in year 2001 and has licensed the LEED Green Building Standard from the USGBC.

Policymakers, planners, and builders can adopt LEED and sustainable building practices as effective strategies to progress towards the UN SDGs. LEED categories contribute to the achievement of the SDGs in the following ways–conserving water, improving energy efficiency, reducing carbon emissions (GHGs), and minimizing harmful air pollutants to enhance environmental sustainability. LEED also supports education, job creation, improved health and well-being, strengthened community resilience, and more. It takes a holistic approach by considering all essential aspects that contribute to the overall quality of a building, rather than solely focusing on individual elements like energy, water, or health, and embraces a comprehensive perspective, recognizing the interplay and synergy among various factors to achieve optimal building performance.

The **primary objective** of LEED is to facilitate the design of good buildings that exude the following characteristics, as outlined by USGBC;

- *Reduce contribution to global climate change.*
- *Enhance individual human health.*

- *Protect and restore water resources.*
- *Protect and enhance biodiversity and ecosystem services.*
- *Promote sustainable and regenerative material cycles.*
- *Enhance community quality of life.*

Certification types and Credit scoring

There are **various kinds of LEED certifications** depending on the types and the phase of the building of projects, as follows;

- ***LEED Building Design and Construction*** **(BD+C)** For new construction or major renovations, and also includes applications for Schools, Retail, Hospitality, Data Centers, Warehouses & Distribution Centers, and Healthcare.
- ***LEED Interior Design and Construction*** **(ID+C)** for complete interior fit-out projects, Includes Commercial Interiors, and also includes applications for Retail and Hospitality.
- ***LEED Building Operations and Maintenance*** **(O+M)** for existing buildings that are undergoing improvement work or little to no construction. Includes Existing Buildings, and also includes applications for Schools, Retail, Hospitality, Data Centers, and Warehouses & Distribution Centers.
- ***LEED Neighbourhood Development*** **(ND)** for new land development projects or redevelopment projects containing residential uses, non-residential uses, or a mix. Projects can be at any stage of the development process, from conceptual planning to construction. Includes Plan and Built Project.
- ***LEED Homes*** for single family homes, low-rise multi-family (one to three stories) or mid-rise multi-family (four or more). Includes Homes, Multifamily Low-rise, Multifamily Midrise.
- ***LEED Cities*** for entire cities and sub-sections of a city. LEED for Cities projects can measure and manage their city's water consumption, energy use, waste, transportation and human experience.

To obtain **LEED certification**, a project accumulates points by adhering to prerequisites and fulfilling criteria related to carbon emissions, energy efficiency, water conservation, waste management, transportation, materials usage, and indoor environmental quality. The certification process involves a thorough examination and assessment by GBCI (Green Business Certification Inc.), which assigns points based on the project's adherence to LEED standards. The total number of points determines the level of LEED certification: *Certified* (40-49 points), *Silver* (50-59 points), *Gold* (60-79 points), or *Platinum* (80+ points). Notably, climate change accounts for 35% relate to climate change, 20% directly impact human health, 15% impact water resources, 10% affect biodiversity, 10% relate to the green economy, and 5% impact community and natural resources. In LEED v4.1, operational and embodied carbon have the most LEED credits associated to them.

CERTIFIED 40-49 points | SILVER 50-59 points | GOLD 60-79 points | PLATINUM 80+ points

LEED Credits and categories

The **major categories and their respective credits** are;

- ***Energy and Atmosphere (33 credits)***
  - Fundamental Commissioning and Verification [*prerequisite*]
  - Minimum Energy Performance [*prerequisite*]
  - Building-Level Energy Metering [*prerequisite*]
  - Fundamental Refrigerant Management [*prerequisite*]
  - Optimize Energy Performance [18]
  - Enhanced Commissioning [6]
  - Advanced Energy Metering [1]
  - Renewable Energy [5]
  - Enhanced Refrigerant Management [1]
  - Grid Harmonization [2]
- ***Location and Transportation (16 credits)***
  - Sensitive Land Protection [1]
  - High Priority Site and Equitable Development [2]
  - Surrounding Density and Diverse Uses [5]
  - Access to Quality Transit [5]
  - Bicycle Facilities [1]
  - Reduced Parking Footprint [1]
  - Electric Vehicles [1]
- ***Materials and Resources (13 credits)***
  - Storage and Collection of Recyclables [*prerequisite*]
  - Building Life-Cycle Impact Reduction [5]
  - Environmental Product Declarations [2]
  - Sourcing of Raw Materials [2]
  - Material Ingredients [2]
  - Construction and Demolition Waste Management [2]
- ***Sustainable Sites (10 credits)***
  - Construction activity pollution prevention [1]
  - Site Assessment [1]
  - Protect or Restore Habitat [2]
  - Open Spaces [1]
  - Rainwater Management [3]
  - Heat Island Reduction [2]
  - Light Pollution Reduction [1]
- ***Water efficiency (10 credits)***
  - Outdoor Water Use Reduction [*prerequisite*]
  - Indoor Water Use Reduction [*prerequisite*]
  - Building-Level Water Metering [*prerequisite*]

- Outdoor Water Use Reduction [2]
- Indoor Water Use Reduction [6]
- Optimize Process Water Use [2]
- Water Metering [1]

- ***Innovation (6 credits)***
  - Innovation in design [5]
  - LEED Accredited Professional [1]
- ***Regional Priority Credits (4 credits)***
- ***Integrative Process (1 credit)***

In the US Green Building (USGBC) annual list for **2021**, **India** achieved the remarkable position of **third in the world for LEED** Green buildings, with a total of 146 certified buildings and spaces, covering an impressive 2.8 million gross area square meters.

Other Green Building Certifications in India

Apart from LEED, the other 2 main green building certification systems in India, are; GRIHA (Green Rating for Integrated Habitat Assessment), India's own rating system jointly developed by TERI and the Ministry of New and Renewable Energy, Government of India, and BEE (Bureau of Energy Efficiency) developed its own rating system for the buildings and the Energy Performance Index (EPI).

GRIHA introduces a comprehensive approach of assessing 11 Parameters with 30 criteria accumulating to 100+5 maximum points (refer Fig. 6), across three stages:

1. ***Pre-construction stage:*** This stage addresses both on-site and off-site factors, such as the proximity to public transportation, soil type, land characteristics, location, existing flora and fauna, and natural landscape features before construction begins.
2. ***Building planning and construction stages***: This stage focuses on resource conservation, reducing resource demand, optimizing resource utilization efficiency, promoting resource recovery and reuse, and ensuring occupant health and well-being.
3. ***Building operation and maintenance stage***: This stage encompasses the operation and maintenance, monitoring and recording energy consumption, ensuring occupant health and well-being, and addressing factors that impact the local and global environment.

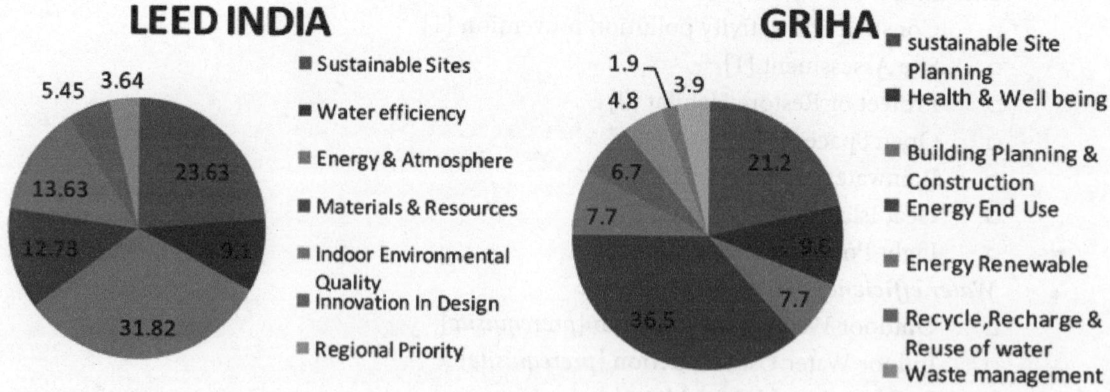

Fig. 5.6: Performance criteria comparison of LEED and GRIHA Rating (Phadtare and Sande, 2015)

Ji et al., 2017, conducted a comparative analysis of GRIHA v2015 and LEED v4 for BD + C: New Construction and Major Renovation with respect to the three pillars of Sustainability and reported;

GRIHA v2015	LEED v4 BD+C
Strong focus on Environmental and Economic aspects, less Social	Has more balanced focus on all the three pillars
Has a dedicated section for assessment of socio-economic aspects of the laborers and service professionals, creating awareness on environment and design for accessibility, missing from LEED.	Has a section dedicated to Regional Priority: Specific Credit, and has social aspects like, open space access, bicycle facilities, diversity in use and quality views included in its assessment. Can be an addition to GRIHA.

GRIHA v.2019			
Section	**Criterion No.**	**Criterion Name**	**Maximum Points**
1. Sustainable Site Planning	1	Green Infrastructure	5
	2	Low Impact Design	5
	3	Design to Mitigate UHIE	2
2. Construction Management	4	Air and Soil Pollution Control	1
	5	Top Soil Preservation	1
	6	Construction Management Practices	2
3. Energy Efficiency	7	Energy Optimization	12
	8	Renewable Energy Utilization	5
	9	Low ODP and GWP Materials	1
4. Occupant Comfort	10	Visual Comfort	4
	11	Thermal and Acoustic Comfort	2
	12	Maintaining Good IAQ	6
5. Water Management	13	Water Demand Reduction	3
	14	Wastewater Treatment	3
	15	Rainwater Management	5
	16	Water Quality and Self-sufficiency	5
6. Solid Waste Management	17	Waste Management-Post Occupancy	4
	18	Organic Waste Treatment On-Site	2
7. Sustainable Building Materials	19	Utilization of Alternative Materials in Building	5
	20	Reduction in GWP through Life Cycle Assessment	5
	21	Alternative Materials for External Site Development	2
8. Life Cycle Costing	22	Life Cycle Cost Analysis	5

9. Socio-Economic Strategies	23	Safety and Sanitation for Construction Workers	1
	24	Universal Accessibility	2
	25	Dedicated Facilities for Service Sta"	2
	26	Positive Social Impact	3
10. Performance Metering and Monitoring	27	Commissioning for Final Rating	7
	28	Smart Metering and Monitoring	0
	29	Operation and Maintenance Protocol	0
Total Points			**100**
11. Innovation	30	Innovation	5
Grand Total Points			**100 + 5**

Fig. 5.7: GRIHA v2019 Criteria and Points

source : www.grihaindigo.org

5.1.5 Intelligent to Smart Buildings

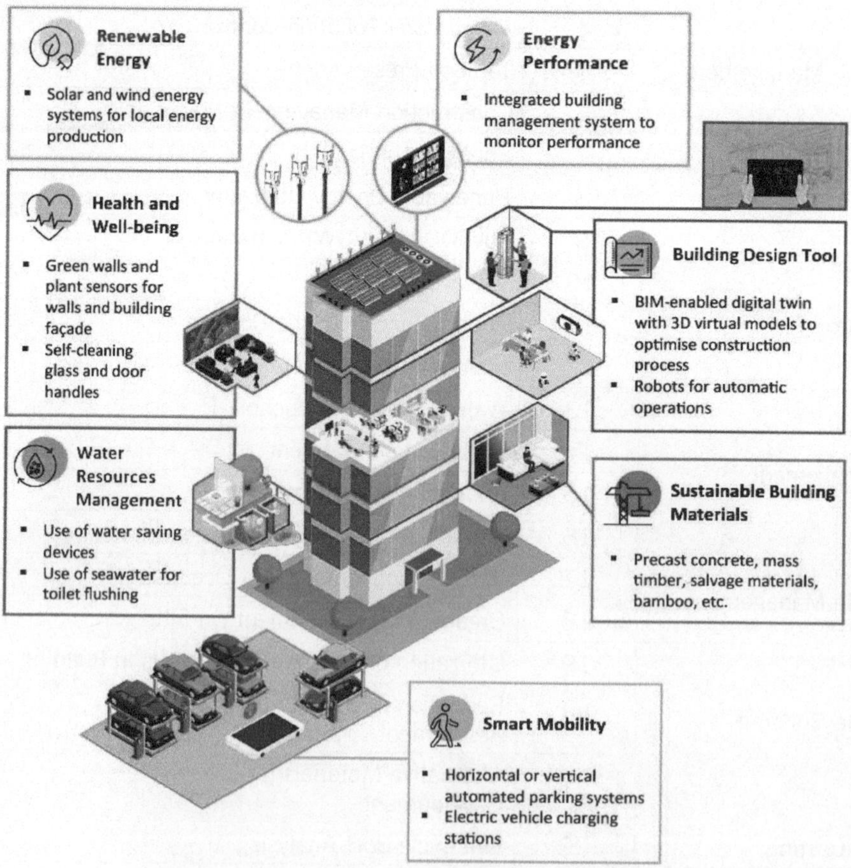

Fig. 5.8: Smart Building technologies and features

The term, 'Intelligent building', is often used commonly to imply and 'Smart building', but their definitions have been evolving with time, since the 1980s when the term was first used to mean "*a building which totally controls its own environment*" (Stubbings, 1988). This equated the building to its BMS and other smart systems, with the intent to reduce human interaction. However, the two terms are complementary, as it may be argued that **Smart Buildings are Intelligent Buildings but with additional, integrated aspects** of '*adaptable control, enterprise and materials and construction*', and addresses both 'intelligence and sustainability issues' to achieve the optimal energy consumption and overall comfort.

Smart Buildings are viewed as a 'subset of smart environments', where smart environments are "*able to acquire and apply knowledge about the environment and its inhabitants in order to improve their experience in that environment*". In a **smart environment**, such as 'Smart City', various elements, i.e., Smart Homes, Smart Buildings, Smart Grids, Smart Meters, Smart Transportation, etc. collaborate synergistically. Smart Buildings prioritize the occupants and **encourage their active participation by incorporating feedback mechanisms, for both receiving and providing information about building usage. These buildings leverage integrated enterprise and intelligent systems to enable inherent control, by empowering occupants to make decisions regarding their comfort, while still maintaining regulated control. Smart buildings promote a personalized and interactive building experience.**

Literature states that Smart Buildings have the **five fundamental** features;

- ***Automation***: the ability to accommodate automatic devices or perform automatic functions.
- ***Multi-functionality***: the ability to allow the performance of more than one function in a building.
- ***Adaptability***: the ability to learn, predict and satisfy the needs of users and the stress from the external environment.
- ***Interactivity***: the ability to allow the interaction among users.
- ***Efficiency***: the ability to provide energy efficiency and save time and costs.

Dakheel, et al. (2020) analysed the above and extended it with four 'Smartness basic features', categorised with respect to four key functions–***Climate response, Grid response***, ***User response*** and ***Monitoring and Supervision***–to include;

- ***Nearly zero energy buildings target*** *(Climate response, Grid response)*
- ***Flexibility*** *(Climate response, Grid response, User response)*
- ***Real-time monitoring*** *(Monitoring and Supervision)*
- ***Real-time Interaction*** *(User response)*

However, every building has a capacity to adapt their operation to the needs of the grid and the occupants, and this is measured by **Smart-readiness indicator** (SRI), developed by EPBD. The SRI in buildings (Verbeke et al., 2017) are particularly important for India as we have stark socio-economic diversity, and introducing the right amount of '*smartness*' holds immense potential for transformative advancements in efficiency, sustainability, occupant comfort, and overall building performance.

5.2 AESTHETICS OF BUILT ENVIRONMENT

Aesthetics, a field within philosophy, is the study of the nature, expression and principles of beauty and artistic taste. The term originates from the Greek word "*aisthetikos*", which refers to sensory perception and understanding, or knowledge acquired through the senses, and hence, varies across people, places and time. There is no one style for aesthetics, but a constant evolution of elements as per the socio-cultural need of that time, or as discussed in earlier Units, as a response to an existing style. It is a deeper, more profound expression of human creativity, meant to be evocative, thought provoking or awe-striking. The **Vitruvian triad** – *firmitas* (durability), *utilitas* (usefulness), and *venustas* (beauty), deemed essential for a good building, captured the significance of aesthetics as **visual appeal encompasses various design elements and principles, such as, its form, size, texture, colour, balance, unity, movement, emphasis, contrast, symmetry, proportion, space, alignment, pattern, as well as cultural relevance and contextual integration.**

However, aesthetics must not always be understood as ornamentation or decoration, but can be grounded in structural integrity and material honesty, as seen in the works of architects like, Frank Lloyd Wright's integration of nature and the use of organic materials in design of Fallingwater, Pennsylvania; or Tadao Ando's Church of the Light in Osaka, Japan, a notable example of honest architecture, where he used concrete as the primary material.

The role of technology–new building materials and construction techniques, also plays an exciting role in developing new aesthetics as a synthesis of technology and art. Italian architect, Pier Luigi Nervi stated, "*the aesthetic sensibility of the designer, who understands their (technology and statistics') intrinsic beauty and validity, welcomes the suggestion and models it, emphasizes it, proportions it in a manner which constitutes the artistic element in architecture*". Parametricism or the parametric design paradigm leverages software like, Rhino with Grasshopper, Fusion 360, Solidworks, etc. to design built forms of unique visual appeal, as championed by architect Zaha Hadid. Therefore, it is crucial to consider not only functionality, safety, serviceability, and durability but also aesthetics, ensuring proper structural performance throughout the entire lifespan.

5.2.1 Role of Urban Arts Commissions

The Delhi Urban Art Commission (DUAC) was established through a parliamentary act in 1973 with the purpose of advising the Government of India on matters related to preserving, developing, and enhancing the aesthetic aspects of urban and environmental design in Delhi. The commission also provides guidance and recommendations to local bodies regarding construction projects, engineering operations, and development proposals that may impact the skyline, aesthetic quality of the surroundings, or public amenities.

The DUAC plays a crucial role in approving, supervising and recommending projects for development and beautification of urban spaces, conservation and beautification of monuments and green landscaped public areas, preservation and maintenance of heritage monuments and buildings, restoration and redevelopment.

It operates in three capacities: as a policy advisor to the government, a regulatory body, and a think tank. In addition, the following **duties** are assigned to the DUAC as per the Act, 1974;

a. Development of district centres, civic centres, areas earmarked for Government administrative buildings and for residential complexes, public parks and public gardens.
b. **Re-development of the area within the jurisdiction of New Delhi Municipal Committee including Connaught Place Complex and its environs, Central Vista, the entire bungalow area of Lutyen's New Delhi, and such other areas as the Central Government may, by notification in the Official Gazette, specify.**
c. Plans, architectural expressions, and visual appearance of new buildings in the centres, areas, parks and gardens specified in clauses (a) and (b) including selections of models or statues and fountains therein.

d. Re-development of areas in the vicinity of Jama Masjid, Red Fort, Qutab, Humayun's Tomb, Old Fort, Tuglakabad and of such other places of historical importance as the Central Government may, by notification in the Official Gazette, specify.
e. Conservation, preservation and beautification of monumental buildings, public parks and public gardens including location or installation of statues or fountains therein.
f. Under-passes, over-passes and regulations of street furniture and hoardings.
g. Location and plans of power houses, water towers, television and other communication towers and other allied structures.
h. any other project or lay-out which is calculated to beautify Delhi or to add to its cultural vitality or to enhance the quality of the surroundings thereof.
i. such other matters as may be prescribed by rules.

5.2.2 Heritage Conservation and Structural Repair and Rehabilitation

Structures, be it built recently or of heritage value, deteriorate due to natural weathering or get damaged due to calamities, historic events, etc. and require design and construction interventions. There are various **types of interventions**, such as, *historic preservation, heritage conservation, restoration, rebuilding, rehabilitation, retro-fitting* etc. and each in turn employs different methods to achieve the desired outcome of reviving the original or intended look and feel of the structure.

Heritage Conservation, objectives and guidelines

Architectural conservation, as a discipline is concerned with the "*integrated informed understanding, sustainable care and appropriate renewal and development of the historic environment*". The aim is to examine the connections between the indoor environment, which is shaped by the interactions among building architecture, materials, structures, services, contents, decorations, and occupants, and the surrounding external environments. It is also the "*process through which the material, historical, and design integrity of any built heritage are prolonged through carefully planned interventions*" (American Conservation Experience). The primary goal is to preserve the original material in its most unchanged state possible, which implies that any repairs or additions made to the built object should not remove, alter, or permanently attach to the original material. It is crucial that all interventions are reversible and removable without causing any damage to the original material, both presently and in the future, and does not involve making artistic choices or experimenting with different materials, i.e., the process of conservation strictly adheres to the object's requirements and specifications, with the sole focus on safeguarding and maintaining the object's authenticity and condition. MoHUA, GoI states that "*Conservation of heritage sites shall include buildings, artifacts, structures, areas and precincts of historic, aesthetic, architectural, cultural or environmentally significant nature (heritage buildings and heritage precincts), natural feature areas of environmental significance or sites of scenic beauty*".

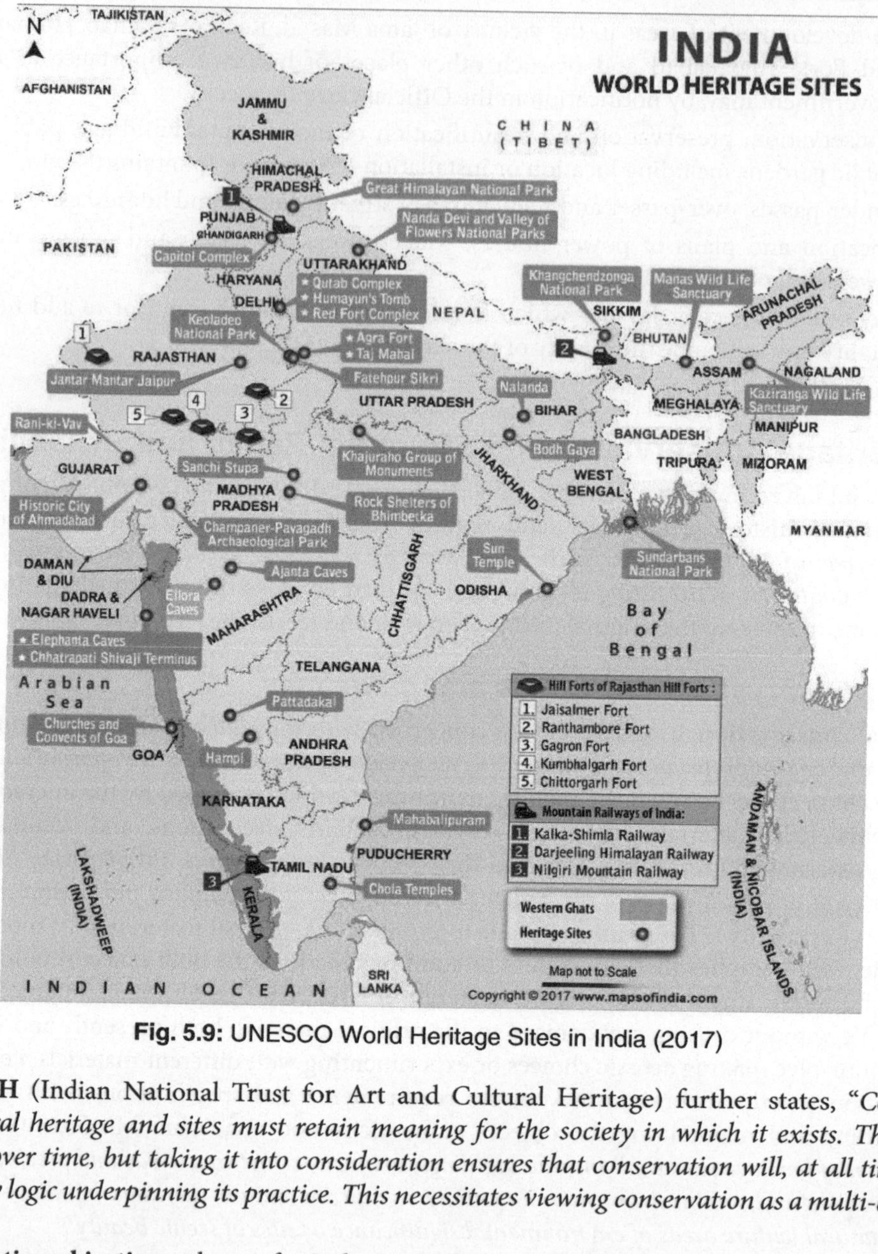

Fig. 5.9: UNESCO World Heritage Sites in India (2017)

INTACH (Indian National Trust for Art and Cultural Heritage) further states, "*Conservation of architectural heritage and sites must retain meaning for the society in which it exists. This meaning may change over time, but taking it into consideration ensures that conservation will, at all times, have a contemporary logic underpinning its practice. This necessitates viewing conservation as a multi-disciplinary activity*".

The **conservation objectives** relevant for India, compiled by INTACH, are as below;

- ***Retain Visual Identity***: Preserving the distinct visual identity of a place, created by its architectural heritage and sites, is crucial amidst global homogenization. This preservation should not mimic legally protected monuments but rather adapt to contemporary society while maintaining heritage relevance. Balancing the needs of heritage and societal changes ensures integration into daily life, while controlling visual clutter like advertisements, cables, and antennas enhances the architectural heritage. Additions, such as, street furniture and signage contribute to appreciating the heritage.

- ***Adaptive Reuse***: Reusing historic buildings and neighbourhoods is economically viable and conserving architectural heritage by involving traditional craftspeople is effective. Prioritizing continuity of original functions, any new use should consider its impact on the local context, adhering to the capacity and vulnerability of the heritage. Comprehensive documentation should accompany any changes to the original fabric, ensuring coherence and engagement with traditional materials and skills. When altering internal functions, the external Fig. should be retained, involving the local community and communicating the benefits of the changes.
- ***Restoration/ Replication/ Rebuilding***: Restoration reinstates the integrity or completeness of architectural heritage/site, aiming to effectively convey its meaning, and may involve reassembly or replacement of missing or deteriorated parts, preceded and followed by documentation. Replication can be appropriate to conserve historic buildings and encourage traditional building methods. Rebuilding at the urban level enhances the visual and experiential quality of the built environment, countering global homogeneity. Reconstruction should be based on minimal physical evidence supported by local crafts people's knowledge and folklore, reinforcing the heritage's bond with society.

  Other sources (URA, Singapore) state that the core principle of conservation that applies to all preserved buildings, regardless of their size or complexity, is based on the ***concept of maximum retention, sensitive restoration, and careful repair,*** often referred to as the "3R"s. Replacement of original elements should only be considered when absolutely necessary, while complete reconstruction contradicts established international conservation practices. When upgrading and adapting a building for new purposes, the existing structure should be preserved by strengthening and repairing its structural elements. Any alterations or reinforcements to these elements should be carried out in a manner that is sympathetic and unobtrusive, utilizing original methods and materials whenever feasible.
- ***Employment Generation***: Conservation strategies should focus on employing local craftsmen, labour, and materials to sustain traditional building practices economically. Utilizing architectural heritage as an alternate strategy meets contemporary needs and supports traditional ways of building.
- ***Local Material and Traditional Technology***: Preference should be given to local materials and traditional technologies, based on available traditional knowledge systems. Modern substitutes should be considered only if proven efficient and judicious, without compromising the integrity of local building traditions. However, caution must be exercised in using certain materials that may damage ecological systems, such as shell lime in coastal areas or wood in general, which may require appropriate substitutions.
- ***Integrated Conservation***: Architectural heritage conservation should be integrated with social and economic aspirations of the community. Conservation-oriented development should be prioritized, necessitating multi-disciplinary teams that include social workers to facilitate dialogue and decision-making, considering diverse social aspirations.
- ***Sustainability***: Conservation should aim to sustain the buildings and/or the traditional skills and knowledge systems for building, positively contributing to the quality of life for the local community over time.

  Several Conservation guidelines are available from INTACH, ICOMOS (international council on monuments and sites), and International Centre for the Study of the Preservation and Restoration of Cultural Property (ICCROM), UNESCO.

Repair and Rehabilitation of Structure

Rehabilitation is the process of restoring the structure to service level, once it had and now lost, strengthening consists in endowing the structure with a service level, higher than that initially planned by modifying the structure not necessarily damaged structure. **Repair** is the process of restoring something that is damaged or deteriorated or broken, to good condition, and is a means to perform rehabilitation through the improvement or modification of a structure, partly or wholly, which is damaged in appearance or serviceability. The primary objective of Repair and Rehabilitation is to *restore the structure, maximizing its functional usefulness*. It also serves the purpose of adapting the structure to fulfil new functional and other requirements. Repair and Rehabilitation techniques are employed for various reasons in structures, including but not limited to the following: damage due to accidents, deterioration due to Environment effects (Fire, Natural calamities like earthquake, Flood, Tsunami, cyclones, Soil and structure interaction, such as, settlement of soil or soil failure), and new functional or loading requirements requiring modifications to structure.

The general approach followed in the **Repair and Rehabilitation** process involves;

- ***Identifying the building*** in need of rehabilitation by performing a Structural Audit of the building.
- ***Gathering information*** about its history, conducting a preliminary survey that includes necessary tests to evaluate various retrofitting options, materials, feasibility and economy.
- ***Identifying the problems*** by performing structural calculations and capacity demand ratio for structural members.
- ***Conceptualising appropriate and feasible solutions***, such as, retrofitting/construction, that aligns with the building's topographical conditions.
- ***Execution of the solution*** by getting the rehabilitation of the building done.
- ***Post repair/retrofitting tests.***

Retrofitting

Retrofitting refers to the engineering process of modifying existing buildings to improve their structural behaviour while preserving their fundamental intended use. It is important to consider the following aspects for retrofitting, as recommended by (Chandar, 2014);

- ***Functionality aspect***: The basic function/ operation of the structure should not be hampered.
- ***Structural safety Aspect***: The susceptibility of the structure to an earthquake event has to be within acceptable standards.
- ***Importance Level Aspect***: Historic buildings with immense archaeological importance are sometimes beyond the cost factor for retrofitting. Such structures have to be rehabilitated without changing its elegance.
- ***Construction Methodology Aspect***: The retrofitting has to be performed using latest construction techniques that have the minimal impact on usual functioning of the buildings.
- ***Economy Aspect***: The entire cost of construction has to be practical and logical towards extended life of the structure.
- ***Skilled labour availability***: The retrofitting practices need unusual construction method and is highly technical job and calls for utmost care to implement it. A very skilled workmanship must be provided to instrument the suggested measures.

Below are the commonly used **techniques for retrofitting** a building:

1. Installation of additional ***Shear Walls*** is usually in the exterior of non-ductile reinforced concrete buildings. The wall maybe cast-in-place or pre-cast.
2. Addition of ***Steel Bracing*** is done for higher strength and stiffness when large openings are required, usually for natural lighting.
3. Application of ***Wall Thickening Technique*** to existing walls of a building is done by adding bricks, concrete and steel reinforcement, and is designed special conditions.
4. Implementation of ***Base Isolation Technique*** of the superstructure from the foundation is done as a passive structural vibration control technique. It buffers seismic loads and protect the building from damage, however, it is inefficient in high-rises.
5. Adoption of ***Mass Reduction Technique,*** such as removal of a storey, is done to decrease the loading.
6. Utilization of ***Jacketing Method*** to strengthen columns and beams by adding a jacket of longitudinal and traverse reinforcement, along with cement, around the structural element.
7. Incorporation of ***Fibre Reinforced Polymer (FRP)*** to axially strengthen columns and enhance its ductility, and in turn, increase load carrying capacity and improve shear capacity of the member.
8. Application of ***Epoxy Injection Method*** to non-moving cracks in concrete walls, slabs, columns, and piers, thereby restoring the strength of the concrete.
9. Implementation of ***External Plate Bonding*** for increasing the shear strength of reinforced concrete beams by completely or partially wrapping steel plates at the joint of a column and beam with external plates or strips, improves the shear strength.
10. ***Section-enlarging Reinforcing Method*** is used to increase the bearing capacity and cross-section stiffness, widely used for RCC beam, slabs, columns.

Fig. 5.10: Retrofitting Solutions (*Top Left to Right*) Shear Wall, Steel Bracing, Section Enlarging reinforcement, (*Middle Left to Right*) FRP Reinforced, (*Bottom Left to Right)* Epoxy Injection Method, External plate Bonding

5.3 INNOVATIONS AND METHODOLOGIES FOR SUSTAINABILITY

Various national and international codes and standards are being developed and updated, grounded on methodologies for optimisation, efficiency and further, integrated with sustainability considerations. These entail for specific civil engineering tasks, design requirements and specifications, construction techniques, environmental management and overall, sustainable infrastructure development.

5.3.1 Building Codes

The International Building Code (IBC), developed by the International Code Council (ICC), is widely used and adopted across the United States and the continent of North America. It also serves as the basis for the legislative building codes for several countries, like Mexico and Abu Dhabi. It focuses on adequate light and ventilation, energy conservation, means of egress facilities, and safety to life and property from fire, explosion, and other hazards; etc. India's indigenous building code, the **National Building Code** is developed by BIS.

The **NBC 2016**, has incorporated a lot of new and salient features (Annexure 2) in keeping with the current automation trends and sustainability goals, a few of which are as follows ;

- Detailed provisions relating to requirements for **accessibility in buildings and built environment** for persons with disabilities and the elderly.
- Norms for **solar energy** utilization.
- **Fire and life safety in modern complex buildings** including the high rises, glazed buildings, atria, commercial kitchen and car parking facilities.
- Updated **structural design provisions for wind and seismic loads**, imposed load due to helipad, and blast loads, for safe design and construction of buildings with due focus on ductile detailing.
- Latest research and development inputs and provisions on concrete, steel and masonry buildings with a view to ensuring **disaster resilient buildings**.
- Updated provisions on engineered use of **bamboo in housing** and other building construction.
- Promotion of use of **agricultural and industrial wastes** including **construction and demolition wastes** in building construction without compromising the quality and safety.
- **New and alternative building materials, and technologies** for building construction such as, reinforced masonry, confined masonry building construction and masonry wall construction using rat-trap bond.
- Inclusion of modern lighting techniques such as **LED and induction light** and their energy consumption.
- Provisions on aviation obstacle lights; **electric vehicle charging** and car park management.
- Use of **refrigerants** for air conditioning addressing zero ozone depletion potential **(ODP)** and ultra-low global warming potential **(GWP)**.
- Inclusion of new and **energy efficient options of air conditioning**, heating and mechanical ventilation, such as variable refrigerant flow system, inverter technology, district cooling system, hybrid central plant using chilled beams, radiant floor components, and geo- thermal cooling and heating.
- Thrust on **envelope optimization** using energy modelling, day lighting simulation, solar shade analysis and wind modelling software to optimize the air conditioning load.

- Air conditioning, heating, and ventilation (HVAC) provisions considering **adaptive thermal comfort conditions** for energy efficiency.
- Updated provisions on **building automation system** to include the latest practices for web-based monitoring and control of performance parameters.
- New chapter on **information and communication enabled installations** in buildings.
- Updated provisions on **rainwater harvesting.**
- New chapter on **solid waste management** covering various solid waste management systems within the building and building complexes.
- Promoting quality of outdoor built environment through updated provisions on **landscape planning**, design and development.
- Promoting **sustainability** in buildings and built environment in tandem with relevant sustainable development goals.
- New chapter on **asset and facility management** to cover provisions relating to management of building assets and associated services, also covering responsibilities of occupants for maintenance of facilities, such as structures, equipment and exterior property.

Another complimentary body is the Indian Society of Heating, Refrigerating and Air- Conditioning Engineers (ISHRAE), which is the only national body in India which has developed **code for IEQ of buildings**, as stated in their 2015 position paper, and has further emphasized on the need for an integrated design approach that blends IEQ and Energy efficiency.

5.3.2 Building Information Modeling Standards

ISO 19650 standard is an internationally recognized guideline for effectively managing information throughout the entire lifecycle of a built asset using building information modeling (BIM). It encompasses the same principles and high-level requirements as the UK BIM Framework and aligns closely with the existing UK 1192 standards.

It is a series of international standards that provide guidelines and requirements for the organization and management of information within the construction industry, including civil engineering projects, by promoting the use of Building Information Modeling (BIM) and standardized processes for information management. It covers various aspects of information management, such as the creation, exchange, and management of digital information throughout the lifecycle of a construction project, with the aim to improve collaboration, efficiency, and interoperability among project stakeholders via BIM. ISO 19650 also addresses topics such as data classification, information security, and documentation requirements. The ISO 19650 standard series includes ;

- **BS EN ISO 19650-1**: This part focuses on the organization and digitalization of information related to buildings and civil engineering works, including building information modelling. It outlines the concepts and principles of information management using BIM.
- **BS EN ISO 19650-2**: This part deals with the organization and digitalization of information during the delivery phase of assets in buildings and civil engineering works. It provides guidance on information management using BIM during this stage.
- **BS EN ISO 19650-3:2020:** This part addresses the organization and digitalization of information in the operational phase of assets in buildings and civil engineering works. It covers information management using BIM during this period.

- **BS EN ISO 19650-5:2020**: This part focuses on the organization and digitalization of information, including building information modelling, with a security-minded approach to information management.

These standards are founded on the UK's standards for information management using building information modelling, namely BS 1192:2007 + A2:2016 and PAS 1192-2:2013. PD 19650-0–UK Transition Guidance, which will along with the UK National Forewords and National Annex aid implementation of the ISO standards in the UK

5.3.3 Environment Management System Standards

An **Environmental Management System** is "*a framework that helps an organization achieve its environmental goals through consistent review, evaluation, and improvement of its environmental performance*" (EPA). It enables organizations to address their regulatory obligations in a systematic and cost-effective manner, proactively reduces environmental risks and enhances health and safety practices for both, occupants, and the public. Additionally, an EMS can tackle non-regulated concerns like, energy conservation, while promoting better operational control and encouraging stakeholder stewardship.

The **fundamental components of an Environmental Management System** encompass;

1. Assessing the organization's environmental objectives.
2. Analysing environmental impacts and compliance obligations, including legal and other requirements.
3. Establishing environmental objectives and targets aimed at reducing environmental impacts and meeting compliance obligations.
4. Implementing programs to achieve these objectives and targets.
5. Monitoring and measuring progress towards the established objectives.
6. Ensuring employees are well-informed and possess the necessary environmental awareness and competence.
7. Conducting regular reviews to track progress and identify opportunities for improvement.

The **ISO 14001:2015** is a standard that establishes the requirements for an Environment Management System to enable organizations to improve their environmental performance, fulfilment of compliance obligations and achievement of environmental objectives. ISO 14001:2015 is applicable to organizations of all sizes, types, and industries, and covers the environmental aspects of an organization's activities, products, and services that the organization can control or influence taking a life cycle perspective into account. It can be utilised either in its entirety or partially, to systematically enhance their environmental management practices, however, it does not establish specific environmental performance criteria.

5.3.4 ISO TC268 'Sustainable Cities and Communities'

TC 268 in the area of **Sustainable Cities and Communities** focuses on standardization efforts that involve developing requirements, frameworks, guidance, and tools to promote sustainable development, with considerations for smartness and resilience. The aim is to support all cities, communities, and stakeholders in rural and urban areas in their journey towards increased sustainability, in line with the SDGs. It has several working groups in the areas of infrastructure metrics, Integration and interaction framework for smart community infrastructures, Data exchange and sharing for smart community infrastructures, Power plant, Disaster risk reduction and Utility tunnel.

ISO/TC268, is responsible for the ISO 37100: 2016 series of standards that assists cities in defining their sustainability objectives and implementing strategies to achieve them. ISO/TR 37150 introduced indicators like Global City Indicators, Green City Index, and Smart City concepts driven by ICT. Efforts are underway to enhance existing indicators, such as, ISO/NP 37122, which focuses on "Sustainable Development in Communities–Indicators for Smart Cities" through legislations (currently in the proposal phase, and soon to be replaced by the ISO/AWI 37100)

The world is constantly developing new knowledge, methodologies, and innovative tools to strategically address improvements in the built environment and strive towards sustainability, due to its profound impact – at the societal and global level. It is imperative that the civil engineer of the future is much more than a master builder, but must gain competence in several other areas of leadership to ensure that the aim of sustainable development is achieved.

UNIT SUMMARY

This elementary unit presents the three pronged story of a good built-environment, as described by Vitruvius – utility, durability and beauty (aesthetics), from the lens of sustainable development. The unit elaborates the various sustainability practices and applications within the domain, such as, Facility management and sustainability strategies, Building control systems, Building certification and rating, and Smartness readiness Index, to enable the civil engineer for sound environmental decision-making. Additionally, the unit further introduces the importance and relevance of 'aesthetics' and its need to preserve the culture inherent in built-environment through practices of heritage conservation, and repair and rehabilitation. Lastly, the knowledge of the various national and international codes and standards for ensuring prioritisation and incorporation of environmental sustainability in practice is propounded.

EXERCISES

I. Multiple Choice Questions

Q.1 Which of the following are not key activities of Facility Management?
   a. Maintenance, cleaning, testing, and inspection
   b. Regulatory Compliance
   c. Heritage Conservation
   d. Safety and Security

Q.2 What are practical strategies of sustainable facility management?
   a. Waste management
   b. Green Building Certification
   c. Performance Monitoring
   d. all of the above

Q.3 Which is not a design strategy for reduction of embodied energy?
   a. Use of recyclable materials
   b. Parametricism
   c. Design for deconstruction
   d. Low maintenance design

Q.4 Which LEED category has the highest associated credit points (33 credits)?
   a. Location and Transportation
   b. Materials and Resources
   c. Energy and Atmosphere
   d. Water efficiency

Q.5 Name India's national body responsible for developing contextually appropriate Heritage Conservation guidelines and objectives?

a. INTACH
b. ICOMOS
c. ICCROM
d. UNESCO

Answers of Multiple Choice Questions				
5.1 **c**	5.2 **d**	5.3 **b**	5.4 **c**	5.5 **a**

II. Short and Long Answer Type Questions

Q.6 What is Aesthetics? Why is it an important consideration in built-environment.

Q.7 What is a Building Control System? Illustrate the different types of sub-systems that it may have and their functionalities.

Q.8 Discuss the various Green Building Certification/Rating systems used in India. Discuss one in detail.

Q.9 What is embodied energy and embodied carbon? What are some of the design strategies for reduction of embodied energy and embodied carbon, elucidate with examples.

Q.10 What are some of the new and innovative additions to the NBC 2016? Explain any two that will have profound impact on environmental sustainability.

KNOW MORE

History of Facility Management

https://www.rentokil-pestcontrolindia.com/facilities-management/history/

About Embodied Carbon

https://worldgbc.org/advancing-net-zero/embodied-carbon/

Environmental Management System

https://www.epa.gov/ems/learn-about-environmental-management-systems#costs

About Smart Growth: What is Built Environment?

https://www.epa.gov/smartgrowth/about-smart-growth#what-is-the-built-environment

REFERENCES AND SUGGESTED READINGS

Books and Publications

[Book] Fennimore, J. P. (2014). *Sustainable Facility Management: Operational Strategies for Today.* Boston, MA, USA: Pearson.

[Book] Roper, K., & Payant, R. (2014). *The facility management handbook.* Amacom

[Book] Nervi, P. (2018). *Aesthetics and Technology in Building: The Twenty-First-Century Edition.* University of Illinois Press.

[Book] Quereshi, S., Jadhao, P. R., Pandey, A., Ahmad, E., & Pant, K. K. (2021). *Sustainable Fuel Technologies Handbook.*

Al Dakheel, J., Del Pero, C., Aste, N., & Leonforte, F. (2020). *Smart buildings features and key performance indicators: A review.* Sustainable Cities and Society, 61, 102328.

Ali, A.M., and Acharya, S. (2023). *Understanding Indoor Environmental Quality (IEQ) of Naturally Ventilated Educational Buildings – A systematic literature review.* In Proceedings of International Conference on Research into Design. Springer, Singapore.

Buckman, A. H., Mayfield, M., & Beck, S. B. (2014). *What is a smart building?* Smart and Sustainable Built Environment, 3(2), 92-109.

Chandar, S. S. (2014). *Rehabilitation of buildings.* International Journal of Civil Engineering Research, 5(4), 333-338.

Gupta, J., & Chakraborty, M. (2021). *Energy efficiency in buildings.* In Sustainable Fuel Technologies Handbook (pp. 457-480). Academic Press.

Heeren, N., Mutel, C.L., Steubing, B., Ostermeyer, Y., Wallbaum, H. and Hellweg, S., 2015. *Environmental Impact of Buildings. What Matters?* Environmental science & technology, 49(16), pp.9832-9841.

Huang, B., Gao, X., Xu, X., Song, J., Geng, Y., Sarkis, J., ... & Nakatani, J. (2020). *A life cycle thinking framework to mitigate the environmental impact of building materials.* One Earth, 3(5), 564-573.

Ji, S., Banerjee, S., & Punekar, R. M. (2017). *Assessment of GRIHA and LEED on the Parameters of Sustainable Design and Development of Buildings.* In Research into Design for Communities, Volume 2: Proceedings of ICoRD 2017 (pp. 179-188). Springer Singapore.

Kralj, D., & MariČ, M. (2008). *Building materials reuse and recycle.* WSEAS transactions on environment and development, 4(5).

Lupíšek, A., Vaculíková, M., Manľík, Š., Hodková, J., & Růžiľka, J. (2015). *Design strategies for low embodied carbon and low embodied energy buildings: principles and examples.* Energy Procedia, 83, 147-156.

Sande, I. I., & Phadtare, N. S. (2015). *Comparative study of LEED and GRIHA rating system.* Journal Of Information, Knowledge And Research, 3(2), 168-174.

World Health Organization (WHO). (2016). *Urban Green Spaces and Health: A Review of Evidence.* Geneva: World Health Organization.

Web sources

https://www.epa.gov/ems

https://www.iso.org/standard/60857.html

https://smart-cities-marketplace.ec.europa.eu/sites/default/files/pimes_guide_for_bioclimatic_design.pdf

https://mohua.gov.in/upload/uploadfiles/files/duaca_act(1).pdf

https://www.duac.org.in/u/Default

https://www.bis.gov.in/standards/technical-department/national-building-code/

https://www.ashrae.org/file%20library/about/position%20documents/pd_indoor-air-quality-2020-07-01.pdf https://ishrae.in/Content/Download/ISHRAE_IEQ_Feb_26_2019_public_draft.pdf

06 CIVIL ENGINEERING PROJECTS

UNIT SPECIFICS

Through this unit we have discussed the following aspects:

- ***Civil Engineering projects and its impact***
  - *Environmental Impact Analysis procedures*
  - *Sustainable Construction–Waste avoidance/ Efficiency increase Advanced construction techniques for better sustainability*
  - *Techniques for reduction of Green House Gas emissions in various aspects of Civil Engineering Projects*
- ***Project Management and Contributions of Civil Engineers***
  - *Paradigms & Systems–Waterfall / Traditional Project Management, Agile Construction Management, and Lean Construction Management*
  - *Quality of products, Health & Safety aspects for stakeholders*
  - *Demand and Contribution of Civil Engineers*
- ***Innovations and methodologies for ensuring Sustainability in Projects***

Besides giving a large number of multiple-choice questions as well as questions of short and long answer types marked in two categories following lower and higher order of Bloom's taxonomy, a list of references and suggested readings are given in the unit so that one can go through them for practice.

There is a "Know More" section, which has been carefully designed so that the supplementary information provided in this part becomes beneficial for the users of the book. It is important to note that for getting more information on various topics of interest some QR codes have been provided which can be scanned for relevant supportive knowledge This section mainly highlights applications of the subject matter for our day-to-day real life or/and industrial applications on variety of aspects, case study related to environmental, sustainability, social and ethical issues whichever applicable, and finally inquisitiveness and curiosity topics of the unit.

RATIONALE

This concluding unit on civil engineering projects, offers the civil engineer a pragmatic view of the tasks and responsibilities to be shouldered during professional practice, be it with respect to environmental sustainability assessment and mitigation on site, or project management aspects.

UNIT OUTCOMES

List of outcomes of this unit is as follows:

U6-O1: Knowledge on Civil Engineering projects and its impact

U6-O2: Knowledge on Project Management and Contribution of Civil Engineering

U6-O3: Knowledge on Innovations and methodologies for ensuring Sustainability during Project development.

Unit-6 Outcomes	EXPECTED MAPPING WITH COURSE OUTCOMES (1- Weak Correlation; 2- Medium correlation; 3- Strong Correlation)						
	CO-1	CO-2	CO-3	CO-4	CO-5	CO-6	CO-7
U6-O1	3	3	3	2	3	3	2
U6-O2	2	1	2	3	3	3	3
U6-O3	1	3	2	1	3	2	1

Civil Engineering projects are multi-faceted and to address its impact on the environment, various procedures, activities, and strategies are employed by large, multidisciplinary teams, sometimes across geographies. By embracing sustainable practices and technologies, civil engineering projects can minimize their ecological footprint and contribute to a more sustainable future.

6.1 CIVIL ENGINEERING PROJECTS AND ITS IMPACT

A construction project is a collaborative effort with "*a group of interrelated work activities constrained by a specific scope, budget, and schedule to deliver capital assets needed to achieve the strategic goals of an Agency*" (Shadan and Fleming, 2012), where the agency or client, maybe an individual, a private or public enterprise.

A construction project typically has the following phases (refer Fig. 1);

- ***Project initiation*** which entails, defining the project requirements, scoping the project, develop a project delivery strategy and contract management planning.
- ***Planning/Environmental Clearance and Real Estate Acquisition*** which entails, planning various site studies, checking compliance and applying for Environmental clearance, acquiring or relocating and taking possession of the land.
- ***Design*** entails designing the built environment as per stakeholder requirements and site conditions, design management, design reviews, feasibility/constructability reviews, risk assessment, Quality assurance and Quality control, incorporating Sustainability (Green building) standards and codes, and Certification.
- ***Construction*** entails construction management, third-party coordination, Quality management, Safety management during construction.
- ***Commissioning*** entails validating the building equipment and systems with operations personnel and third parties.
- ***Project Close-out / Operations*** is when the project begins to function as intended and all hand-over is done to the Agency/client.

The success of a project requires; Collaborating with architects, engineers and contractors to develop plans and establish timelines, Estimating and negotiating project costs, Creating and monitoring project budgets, Securing permits and design evaluations, Making schedules and work timetables, Determining what methods and strategies are appropriate for a project, Communicating with clients, contractors and other stakeholders, Assembling and leading construction teams, Working with building, construction and regulatory specialists, and Managing the day-to-day workflow of a project. The environmental impact of the project has to be systematically addressed and mitigated at each phase through procedure and a sustainable construction commitment.

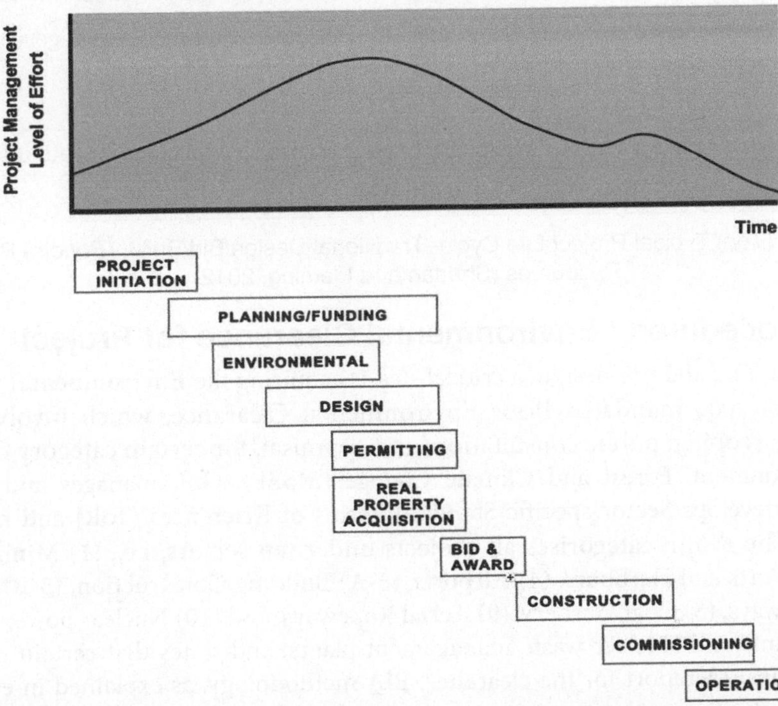

Capability	Resource	Function
Management and Control	Project Management	Manage the overall project and its phases -initiation, planning, design, construction, and closeout
	Project Management Oversight	Oversee project performance by a party independent of the project team and report to project sponsor(s)
	Configuration Management	Control changes to project deliverables/scope of work
	Cost Control	Control project costs within budget
	Schedule Control	Control project progress within schedule
	Accounting	Record project expenditures, issue payments, and manage project funding
	Records Management	Capture, store, control, and retrieve project records/documents
	Procurement/ Administration	Procure and administer project contracts

Capability	Resource	Function
Planning, Engineering, and Technical	Architectural	Develop architectural and design concepts
	Engineering	Prepare detailed engineering and final design documents
	Environmental	Undertake environmental planning and clearance
	Real Estate	Acquire real estate and perform relocation functions
	Communications	Communicate with the community/media/ government
Construction and Supply	Construction Contractor	Construct facilities
	Third Party Agency	Relocate or gain access to public and private utilities
	Equipment Supplier	Supply/install equipment

Fig. 6.1: (*Top*) Typical Project Life Cycle–Traditional Design/Bid/Build, (*Bottom*) Project Resources (Shadan and Fleming, 2012)

6.1.1 EIA Procedures : Environmental Clearance for Project

During project planning and pre-design, a critical step is acquiring the **Environmental Clearance**. The EIA Notification of 2006 mandated Prior Environmental Clearance, which involves four stages namely, screening; scoping; public consultation; and appraisal, for certain category of projects. The Ministry of Environment, Forest and Climate Change (MoEF), GoI, manages and publishes EIA Notifications and develops Sector specific Standard Terms of References (ToR) and manuals for the different sectors. The MoEF categorises all projects under ten sectors, i.e., (1) Mining, (2) Mineral Beneficiation, (3) Ports and Harbours, (4) Airports, (5-A) Building Construction, (5-B) Townships, (6) Asbestos, (7) Highways, (8) Coal Washery, (9) Aerial Ropeways, and (10) Nuclear power plants, Nuclear fuel processing plants and Nuclear waste management plants, and states that certain type (B) do not require to submit an EIA report for the clearance. EIA methodology, as explained in earlier Unit, has various steps, namely: Screening, Scoping, Impact Analysis, Mitigation, Reporting, Review of EIA, Decision-Making, and Post Monitoring, to develop a report on the state of affairs of the proposed project and its possible impact on the environment.

To begin an **Environmental Clearance for Project,** following are the steps outlined by the Centre for Science and Environment, India;

Step 1: Project proponent identifies the location of proposed plant after ensuring compliance with existing siting guidelines.

Step 2: The project proponent then assesses if the proposed activity/project falls under the purview of environmental clearance. If it is mentioned in schedule of the notification, the proponent conducts an EIA study either directly or through a consultant, however, B2 projects do not require preparation of EIA reports.

Step 3: After the EIA report is ready, the investor approaches the concerned State Pollution Control Board (SPCB) and the State Forest Department (if the location involves use of forestland). The SPCB evaluates and assesses the quantity and quality of effluents likely to be generated by the proposed unit as well as the efficacy of the control measures proposed by the investor to meet the prescribed standards. If the SPCB is satisfied that the proposed unit will meet all the prescribed effluent and emissions standards, it issues consent to establish (popularly known as NOC), which is valid for 15 years.

Step 4: The process of public hearing is conducted prior to the issue of NOC from SPCB. The District Collector is the chairperson of the public hearing committee. Other members of the committee include the official from the district development body, SPCB, Department of Environment and Forest, Taluka and Gram Panchayat representative, and senior citizen of the district, etc. The hearing committee hears the objections/suggestions from the public and after inserting certain clauses it is passed on to the next stage of approval (Ministry of Forest and Environment).

Step 5: The project proponent **submits an application for environmental clearance** with the MoEF if it falls under Project A category or the state government if it falls under project B category. The application form (1 and 1A, *details given below in next Section*) is submitted with EIA report, EMP, details of public hearing and NOC granted by the state regulators.

Step 6: Environmental appraisal: The documents submitted by an investor are first scrutinised by a multi-disciplinary staff functioning in the Ministry of Environment and Forests who may also undertake site-visits wherever required, interact with the investors and hold consultations with experts on specific issues as and when necessary. After this preliminary scrutiny, the proposals are placed before specially constituted committees of experts whose composition is specified in the EIA Notification.

Step 7: *Issues of clearance or rejection letter:* When a project requires both environmental clearance as well as approval under the Forest (Conservation) Act, 1980. The clearance granted shall be valid for a period of five years for commencements of the construction or operation of the project.

Application Forms for Environmental Clearance

Details of **Form1** is to be filled, which comprises of general project information and a checklist for confirmation on the Activity and Environmental Sensitivity of the proposed site is to be filled out, as below;

1. ***Basic Information,*** *on*
   1.1. Name of project, Proposed capacity/area, Type of Project (New, Modernisation, Expansion), Existing capacity/area, Category of project, Location, etc.
   1.2. Whether the proposal involves approval/clearance under: if, yes details of the same and their status to be given a) The Forest (Conservation) Act, 1980? (b) The wildlife (protection) Act, 1972? (c) The CRZ Notification, 1991?
   1.3. Whether there is any Government Order/policy relevant/relating to the site?
   1.4. Forest land involved (hectares)
   1.5. Whether there is any litigation pending against the project and/or land in which the project is proposed to be set up?
2. ***Activity*** *(Each sub-section has several Information/Checklist confirmation)*
   2.1. Construction, operation or decommissioning of the Project involving actions, which will cause physical changes in the locality (topography, land use, changes in water bodies, etc.)
   2.2. Use of Natural resources for construction or operation of the Project (such as land, water, materials or energy, especially any resources which are non-renewable or in short supply):
   2.3. Use, storage, transport, handling or production of substances or materials, which could be harmful to human health or the environment or raise concerns about actual or perceived risks to human health.
   2.1. Production of solid wastes during construction or operation or decommissioning (MT/month)

2.2. Release of pollutants or any hazardous, toxic or noxious substances to air (Kg/hr)

2.3. Generation of Noise and Vibration, and Emissions of Light and Heat:

2.4. Risks of contamination of land or water from releases of pollutants into the ground or into sewers, surface waters, groundwater, coastal waters or the sea:

2.5. Risk of accidents during construction or operation of the Project, which could affect human health or the environment.

2.6. Factors which should be considered (such as consequential development) which could **lead to environmental effects or the potential for cumulative impacts with other existing or planned activities in the locality.**

3. ***Environmental Sensitivity***

3.1. Areas protected under international conventions, national or local legislation for their ecological, landscape, cultural or other related value

3.2. Areas which are important or sensitive for ecological reasons–Wetlands, watercourses or other water bodies, coastal zone, biospheres, mountains, forests

3.3. Areas used by protected, important or sensitive flora or breeding, foraging, resting, over wintering, migration.

3.4. Inland, coastal, marine or underground waters

3.5. State, National boundaries

3.6. Routes or facilities used by the public for access to recreation or another tourist, pilgrim areas

3.7. Defence installations

3.8. Densely populated or built-up area

3.9. Areas occupied by sensitive man-made land uses (hospitals, schools, places of worship, community facilities)

3.10. Areas containing important, high quality or scarce resources (ground water resources, surface resources, forestry, agriculture, fisheries, tourism, minerals)

3.11. Areas already subjected to pollution or environmental damage. (*Those where existing legal environmental standards are exceeded*)

3.12. Areas susceptible to natural hazard which could cause the project to present environmental problems (earthquakes, subsidence, landslides, erosion, flooding or extreme or adverse climatic conditions)

Then, Proposed **Terms of Reference** (TOR) for EIA studies applicable, is to be agreed by the proposer.

Details of **Form 1A**, which is a check list of environmental impacts is to be filled out, which is a table of 9 sections across the different environmental categories, which are;

1. Land Environment (1.1-1.9 questions)
2. Water environment (2.1-2.14 questions)
3. Vegetation (3.1- 3.3 questions)
4. Fauna (4.1- 4.3 questions)
5. Air environment (5.1- 5.6 questions)
6. Aesthetics (6.1- 6.4 questions)
7. Socio-economic aspects (7.1- 7.3 questions)
8. Building materials (8.1- 8.4 questions)

9. Energy conservation (9.1- 9.13 questions)
10. Environment management plan (for Construction phase and Functional /Use phase) across 5 subheadings; i.e., Environmental Components (Ambient Air, noise, water, Land Aesthetics), Predicted Impact, Probable source of Impact, Mitigation Measures and Remarks.

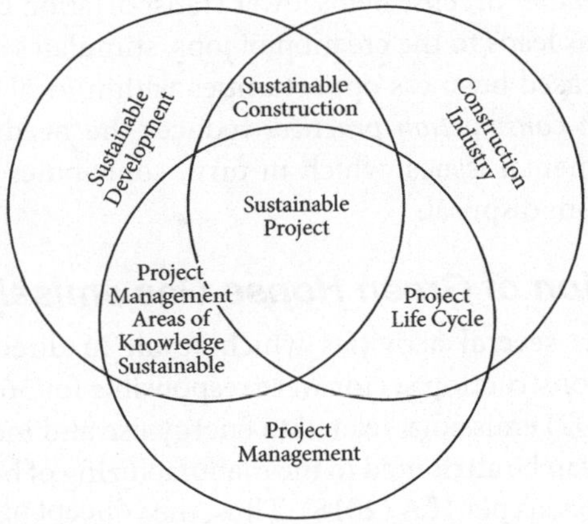

Fig. 6.2: Elements of a Sustainable Project

6.1.2 Sustainable Construction

Sustainable construction is defined as "*how the construction industry together with its product the 'built environment', among many sectors of the economy and human activity, can contribute to the sustainability of the earth including its human and non-human inhabitants*" (Kibert, 2007). The goal of sustainable construction is to minimize the environmental impact caused by the construction industry, through the following objectives:

1. Utilise renewable and recyclable materials.
2. Decrease the embodied energy within building materials.
3. Reduce the energy consumption of the completed building.
4. Minimize on-site waste generation.
5. Safeguard natural habitats.

Sustainable construction encompasses various activities, ranging from extracting materials and manufacturing products, to assembling them into buildings, maintaining and replacing systems, and ultimately disposing of waste, building systems, and the building structure, whilst considering the above mentioned objectives throughout the entire lifecycle of the construction process and the resulting built environment. Additionally, the physical distribution and relationships between buildings and infrastructure play a role in determining consumption patterns, which are influenced by planning decisions. The implementation of sustainable construction is further complicated by the involvement of public policy in the form of regulations, incentives, and disincentives. It also involves the participation of industries such as real estate, finance, and insurance, as well as institutions like higher education, design firms, and construction companies.

Waste Avoidance/ Efficiency Increase

Overall **efficiency increases**, in terms of material usage, manpower and equipment utilisation, energy and other resource consumption, and its associated economic costs, by waste avoidance. *Reducing* *disposal of waste* construction and demolition materials, reduces the environmental impact caused by the extraction and consumption of virgin resources and the production of new materials is offset. It brings about cost reductions in overall building projects by avoiding purchase costs through the reuse of materials and by donating recovered materials to qualified charities, which can provide tax benefits. Furthermore, the *use of onsite material reuse* helps lower transportation costs. Employing *deconstruction and selective demolition methods* leads to the creation of jobs, stimulates economic activities in recycling industries, and generates increased business opportunities within local communities. Last but not the least, *implementing sustainable construction practices* reduces the need for disposal facilities, thereby mitigating associated environmental issues, which in turn, contributes to the conservation of landfill space by diverting materials from disposal.

Techniques for reduction of Green House Gas emissions

The construction industry has several activities which result in direct or indirect Greenhouse gas emissions. The buildings and construction sector were responsible for 36% of total energy consumption and 39% of carbon dioxide (CO2) emissions related to energy use and industrial processes. Out of these emissions, approximately 11% can be attributed to the manufacturing of building materials and products, including steel, cement, and glass, as per IEA (2018). Thus, the concept of embodied carbon captures the implied CO_2 equivalent/GHG emission, as discussed earlier, and is the critical decider when it comes to strategizing techniques to reduce GHG emission.

Techniques recommended by AIA are;

- ***Limit carbon-intensive materials,*** like aluminium, plastics, and foam insulation, and use judiciously.
- ***Choose lower carbon alternatives***, such as, wood structure instead of steel and concrete, or wood siding instead of vinyl, and review Environmental Product Declarations for selecting alternatives.
- ***Choose carbon sequestering materials,*** like wood, straw or hemp insulation and bring
- down the embodied carbon in a project.
- ***Reuse materials***, like brick, metals, broken concrete, or wood, as salvaged materials typically have a much lower embodied carbon footprint than those newly manufactured.
- ***Use high-recycled content materials***, particularly for metals as they are carbon-
- intensive but can be recycled, which brings down its embodied carbon value.
- ***Use fewer finish materials***, such as, polished concrete slabs in place of tiled, carpet or vinyl finished flooring saves the embodied carbon.
- ***Minimize waste***, as discussed above. Modularity and standard sizes for common materials like plywood, gypsum boards, wood framing, and pre-cut structural members can be factored in during design and wastage can be minimised.
- ***Maximize structural efficiency***, as it is the highest contributor to the embodied carbon
- of the project, and using optimum value engineering wood framing methods, efficient structural sections, and slabs are all effective methods to maximize efficiency and minimize material use.
- ***Reuse buildings instead of constructing new ones,*** as renovation and reuse projects
- typically save between 50 and 75 percent of the embodied carbon emissions compared to constructing a new building.

Advanced Construction Techniques For Better Sustainability

Following are the technologies in construction have potential to significantly enhance sustainability;

1. ***Prefabrication*** involves the manufacturing of building elements/components or modules off-site and their assembly on-site, allowing for better quality control, reduced material waste, and faster construction timelines. This technique improves construction efficiency, reduces waste, enhances quality control, and minimizes disruption to the surrounding environment. It also enables easier deconstruction and material reuse at the end of a building's life cycle. In addition, employing a controlled environment in which to construct and pre-fabricate, results in improved quality of building components and less waste, as external conditions can no longer hinder its properties.

Fig. 6.3: Prefabricated building elements

2. ***Modular construction*** takes this a step further by assembling pre-made modules to create complete structures, offering flexibility, reduced on-site disruptions, and potential for deconstruction and reuse.
3. ***Mass Timber Construction*** refers to the use of large, prefabricated timber panels or components for structural elements, such as cross-laminated timber (CLT) and glued- laminated timber (glulam) which offer high strength, reduced carbon emissions, and faster construction times.
4. ***Robotic construction* is an advanced technique involving robots and automation technologies, to perform various construction tasks, such as bricklaying, concrete pouring, and material handling, revolutionizing the construction industry. It offers high accuracy and efficiency, improved safety, enhanced productivity, and reduced labour costs.**
5. ***3D printing*** utilizes robotic arms or gantries to precisely deposit layers of construction materials, such as concrete or specialized composite materials, based on complex and customised digital designs. 3D printing in construction offers several benefits, including faster construction timelines, reduced material waste, enhanced design flexibility, and cost savings, and additionally, allows for the integration of sustainable features, such as incorporating insulation directly into the printed structure.

Fig. 6.4: Advanced technologies : (*Top Left*) Modular construction, (*Top Right*) Mass timber construction, (*Bottom Left*) Robotic construction and (*Bottom Right*) 3D printing

6.2 PROJECT MANAGEMENT

Project management in construction projects refers to the *discipline of planning, organizing, and controlling resources and activities to successfully deliver construction projects within defined constraints of time, budget, and quality.* It involves coordinating various stakeholders, managing risks, and ensuring the project progresses smoothly from initiation to completion The PMC (project management consultant) includes the project manager, project & planning engineer, construction manager, site engineer, surveyor, Quality control engineer, Health and safety officer, etc. It is their responsibility to ensure timely and quality construction of projects.

6.2.1 Project Management paradigms and systems

Paradigms and Approaches

In construction project management, different **paradigms or approaches** are employed based on the project's unique characteristics and requirements, such as;

1. ***Traditional/Waterfall Paradigm***: The traditional paradigm follows a linear sequence of project phases, where each phase is completed before moving on to the next. This includes distinct stages such as initiation, planning, design, construction, and handover. It emphasizes detailed planning, documentation, and a hierarchical management structure.
2. ***Agile Paradigm***: Agile project management focuses on flexibility, adaptability, and iterative development. It emphasizes collaboration, frequent feedback, and continuous improvement. Agile methodologies, such as Scrum and Kanban, allow for dynamic project planning, quick decision-making, and shorter development cycles.

3. ***Lean Construction Paradigm***: Lean construction aims to maximize value and minimize waste by streamlining processes and eliminating non-value-adding activities. It emphasizes continuous improvement, visual management, and the reduction of bottlenecks. Lean principles, derived from the manufacturing industry, are applied to improve efficiency and productivity in construction projects.

Other systemic approaches based on Project Delivery strategy are ;

1. **Design/Bid/Build** approach is when the owner manages the project, contracts out design to engineering consultants and construction to contractors, and retains a General Engineering Consultant (GEC), Construction Management (CM) and Program Management consultant (PMC).
2. **Design/Build** approach is when a single entity, typically a design-build contractor, is responsible for both the design and construction of the project. This integrated approach reduces coordination issues and promotes efficient communication, resulting in faster project delivery and potentially lower costs.
3. **Turnkey or D/B/O/M approach** is when a turnkey contractor is in charge from design conceptualisation to operation and maintenance.
4. **Integrated Project Delivery** (IPD) is a collaborative approach where all project stakeholders, including the owner, architects, engineers, contractors, and suppliers, work together from the project's inception. It emphasizes early involvement, shared decision- making, and risk/ reward sharing, fostering collaboration, and improving project outcomes.
5. **Public-Private Partnership** (PPP) is a contractual arrangement between a public agency and a private sector entity. It combines public sector requirements with private sector innovation and expertise. PPPs can help finance, develop, and operate infrastructure projects, leveraging private sector resources while sharing risks and rewards.

Project Management Systems

Cleland (1977) proposed a model for project management system, through the perspective of '*systems approach*', with the project team at the focus interacting with the various functional sub-systems, as below;

- The ***Organizational Facilitative Subsystem*** refers to the arrangement within an organization that combines project teams with the functional structure. This creates a "matrix" organization that establishes formal authority, responsibility patterns, and reporting relationships to facilitate the initiation and completion of specific projects. In this context, two key organizational units emerge: the project team and the functional units.
- The ***Project Planning Subsystem*** focuses on the selection of projects, identification of project objectives and goals, and the formulation of a strategy to achieve those objectives and goals. Project plans outline the necessary resources and allocation methods to support the project, drawing from the organization's resources regardless of their location.
- The ***Project Control Subsystem*** involves setting performance standards for the project's schedule, budget, and technical aspects. This subsystem incorporates feedback mechanisms to compare actual progress with planned progress and initiates corrective action when necessary. The control subsystem ensures effective monitoring of the various organizational units involved in the project, ensuring timely and within-budget project delivery.

Fig. 6.5: Project management Model, showing the focal position of the project team and the interacting sub-systems (Cleland, 1977)

- The ***Project Management Information Subsystem*** encompasses the essential intelligence required for effective project control. This subsystem can be informal, involving periodic meetings where project participants report on their project work's status. Alternatively, it can be a formal information retrieval system that provides regular updates on project activities. This subsystem enables project team members to make informed decisions and implement effective project management strategies.
- ***Techniques and Methodology***, while not a subsystem in the traditional sense, encompass various management science techniques such as PERT, CPM, PERT-Cost related scheduling techniques, modeling, simulation, linear programming, and regression analysis. These techniques help evaluate risk and uncertainty factors in project decision-making.
- The ***Cultural Ambience Subsystem*** reflects the organization's environment and the practice of project management within it. The cultural ambience encompasses how individuals and social groups perceive and feel about project management practices in the organization. Factors such as emotions, attitudes, assumptions, experiences, and values shape the organization's cultural ambience. This ambience influences individual behaviours, thoughts, feelings, and expressions, ultimately determining socially acceptable behaviour within the organization.

Construction Project Management Software

The Global construction management software market is expected to expand at a CAGR (Compound Annual Growth Rate) of 8.70% from 2020 to 2027 and reach a net worth of $2.73 billion by 2027, as per the (Data Bridge Market Research).

Construction project management software is a solution employed by professionals to streamline the construction planning process. These tools provide engineering estimates and capital maintenance capabilities, making them suitable for managing projects of varying sizes and complexities. It enables users to define categories or jobs, track itemized costs, and automatically generate financial reports. By incorporating features such as GIS/ESRI mapping, data warehouses, and budget and risk analysis, construction project management software facilitates smoother project development processes. Many construction ERP software packages include construction project management as a prominent feature.

Construction software aids in tracking project progress and monitoring offsite teams.

- ***Automated accounting*** functions streamline tasks such as accounts payables/receivables, payroll management, and work order entries.
- ***Job costing*** capabilities within construction management software enable builders and contractors to define costs associated with employees, production managers, contractors, and supervisors.
- ***Service management*** features in top-notch construction management software handle production schedules, work orders, and asset allocation.
- ***Scheduling*** for the project delivery lifecycle is critical, the tool helps with rescheduling timelines, managing resources, and updating schedules.
- ***Workload automation*** capabilities are beneficial for planning project timelines and balancing workloads across multiple channels, including predicting equipment and labor interdependencies.
- ***Reporting functionalities*** in construction project management software provide data summaries, evaluate project status, and present architectural services costs.
- ***Inventory management*** features in construction software help locate lost inventories, track assets on job sites, and conduct inventory audits.

There are several **construction project management software** presently available, each with certain pros and cons, briefly discussed below (https://project-management.com/) ;

Smartsheet provides a collaborative platform that enables users to work together, gather project requirements, and monitor initial cost estimates using a spreadsheet format. Teams can centralize project requirements and conveniently manage documentation by attaching files. The software also offers dashboards and portals to showcase key visual information relevant to the project. Additionally, users can utilize features such as project cost calculations, forms, and reports to enhance project management capabilities.

monday.com is a versatile platform designed to cater to the needs of architects, builders, general contractors, and engineers. It facilitates seamless collaboration and file sharing, enabling users to easily exchange files, images, updates, RFIs (Requests for Information), and feedback. The platform includes a built-in progress tracking feature that automatically monitors, calculates, and updates crucial information such as budgets, timelines, and resource allocation at every stage of the project. This data is presented through high-level dashboards and charts, providing users with a comprehensive overview of project performance. In addition, monday.com offers a range of valuable features including pre-built construction templates, tools for financial management, portfolio management capabilities, and efficient document management.

CoConstruct (Buildertrend) offers streamlined construction project management that incorporates a Schedule feature with Gantt charts and various views to ensure the smooth progress of all jobs. Users have the ability to create, assign, and monitor to-do tasks for team members and subcontractors, and can even set reminders for clients. The software allows for easy creation of to-dos through voice-to-text functionality, along with the ability to attach documents and photos and send automated notifications. Additionally, CoConstruct provides built-in communication tools for clients and trade partners, financial features for estimating and forecasting, capabilities for bidding and proposals, timesheet management, and seamless integration with other systems.

Fieldwire offers a communication platform that enables team members to engage in real-time conversations tied to specific tasks, promoting quicker decision-making and issue resolution. Users have the ability to monitor and manage all construction tasks, activities, and issues within the software. They can document and annotate issues, as well as update plans with detailed notes, photos, and videos. The construction project management software also provides scheduling and reporting features, along with mobile applications for on-the-go access. Field teams can effortlessly update plans and generate digital forms, streamlining the construction process.

Procore construction project management software incorporates mobile collaboration tools that foster task clarity, enabling teams to adhere to schedules and minimize rework. It captures and centralizes all project correspondence, ensuring easy accessibility. Real-time updates of information enhance visibility and mitigate risks. The software offers a comprehensive project overview, tracks all project steps, and expedites the approval process through features like a contact directory, document management, specification management, integrated scheduling, task management, Requests for Information (RFIs), drawing management, time card tracking, and email integration.

6.2.2 Quality of Products, Health and Safety aspects for stakeholder

Quality control and safety in construction projects are critical determinants of project success, and are intertwined with each other as a lack of quality may lead to a safety issue. Defects or failures in constructed facilities can lead to requirement for re-construction, thereby, affecting facility operations and causing project delays, cumulatively resulting in increased costs. It may also lead to accidents causing injury or fatality leading to further costs, such as incentive, insurance, inspection, etc. Specifying **quality requirements** in the design and contract documentation is crucial during the construction process. They should be well-defined and verifiable to ensure understanding and compliance among all project stakeholders. Decisions made during the planning and design phases have implications on both, quality and safety aspects as certain conformity and compliance can be built into the project management process from the start. However, both are subject to site conditions and other uncertainties, such as, change in the client request, or increase in material prices leading to the need of changing designs or revisiting decisions. Effective project managers strive to ensure that the work is done correctly from the beginning and that major accidents, at product and people level, are avoided. Beyond design decisions and planning vigilance, safety depends on education, alertness and cooperation amongst all stakeholders during the construction process.

Bugalia, Maemura and Ozawa (2019) note across their study comparing high-speed railways in India and Japan, that ***safety is a culture*** and that "*that top management must adopt a multi- pronged approach to improve the safety culture of an organization. There is no one- dimensional management strategy that is sufficient for improving the level of an organization's safety culture*". They further report that organisational reforms are required to bring about a shift in the safety culture and may be achieved through strong training systems, and recommend the need to study cultural aspects in tandem with technology, people and organisational aspects.

National and international bodies, such as, **National Occupational Safety and Health (OSH)** in India and Occupational Safety and Health Administration (OSHA) in the US, routinely conducts site visits of workplaces in conjunction with approved state inspection agencies. In India, Health and Safety is overlooked as per the '*The Building and other construction workers (Regulation of employment and conditions of* *service) Act 1996'*, under which, Section 38 deals with Safety Committee and safety officers in every establishment; Section 39 deals with Notice of certain accidents (which causes death or any bodily injury by reason of which the person injured is prevented from working for a period of forty-eight hours or more) and prescriptions for the same; and Section 40 deals with Power of appropriate Government to make rules for the safety and health of building and other construction workers.

6.2.3 Demand and Contribution of Civil Engineering

According to the RICS (Royal Institution of Chartered Surveyors) Research on '*Real Estate and Construction Professionals in India by 2020*', industry projections, India's construction industry is predicted to reach a value of USD 1 trillion by 2030 and contribute around 13% to the country's GDP by 2025. It is the fastest growing sector and is on track to become the largest employer by 2022, providing jobs for over 75 million individuals. Presently, skilled (Supervisors, Technician, Foremen, Tradesmen, etc), semi-skilled and unskilled workers (helpers and Labourers) constitute 95.3% of the workforce, while only 1.2% is occupied by the Core Professionals, such as, Civil Engineers, Architects and Planners.

To fulfil the country's real estate needs (approximately 1.27 million civil engineers) and infrastructure requirements (approximately 3 million civil engineers), an annual average of 4 million civil engineers in the next decade is necessary, in spite of 1.5 million graduates in India every year. Due to the sustained shortage in annual supply and an increasing year-on-year demand, the cumulative demand for civil engineers between 2010 and 2020 is projected to be around 40.2 million. Unfortunately, there is an estimated shortfall of approximately 39.4 million civil engineers during the same period and recent surveys indicate a 6.27 percent decline in employment within the construction industry.

To Civil engineers play a **crucial role** in society by contributing to the development and maintenance of social infrastructure, as well as working towards building a sustainable future.

One of the key contributions of civil engineers is the planning, design, and construction of essential infrastructure such as roads, bridges, airports, water supply systems, and wastewater treatment plants. These infrastructure projects not only facilitate the movement of people and goods but also improve the overall quality of life by providing access to basic amenities and services. Civil engineers ensure that these structures are safe, efficient, and environmentally sustainable, taking into account factors such as traffic flow, environmental impact, and resource efficiency.

Civil engineers also play a vital role in *addressing the challenges of urbanization* and population growth. They are involved in urban planning and the design of sustainable cities, taking into consideration factors like land use, transportation systems, and the efficient utilization of resources. By *integrating principles of sustainable development* into their projects, **civil engineers contribute to reducing environmental impact**, promoting energy efficiency, and creating resilient and liveable urban environments.

In the face of climate change, civil engineers are at the forefront of developing solutions to mitigate its effects and enhance resilience. They work on projects related to flood management, coastal protection, and the design of structures that can withstand natural disasters such as earthquakes and hurricanes. By *incorporating climate adaptation strategies* and utilizing innovative technologies, **civil engineers contribute to building a more resilient society** that can withstand the challenges posed by a changing climate.

Furthermore, **civil engineers contribute to the sustainable use of resources** through their focus on efficient design and construction practices. They promote the use of renewable materials, energy-efficient systems, and sustainable construction techniques, thereby minimizing the environmental impact of infrastructure projects. By embracing concepts such as *green building design and sustainable* *construction practices*, civil engineers help to reduce carbon emissions, conserve resources, and create environmentally friendly structures.

Moreover, **civil engineers actively engage in research and development** to drive innovation in their field. They explore *new materials, technologies, and methodologies* that can improve the efficiency, durability, and sustainability of infrastructure.

6.3 INNOVATIONS AND METHODOLOGIES FOR SUSTAINABILITY

While in the previous Units several technological and methodological innovations have been discussed, for project management a cutting-edge innovation is **Virtual Design and Construction (VDC)**. It is a technology that aids in the coordinated management of comprehensive performance models for design-construction projects, which encompasses various aspects such as the facilities themselves, work processes, and the organization of the design-construction-operation team. VDC involves using digital tools, such as Building Information Modeling (BIM), to visualize, simulate, and coordinate various aspects of a construction project. It enables better collaboration, clash detection, and data-driven decision- making, leading to improved project coordination, cost control, and reduced rework.

VDC offers several benefits, such as;

- **Integrated Approach with a Common Data Environment** (CDE) and allowing communication and collaboration with different project team members, third party members and other stakeholders.
- **Risk mitigation and Enhanced safety** for workers and end-users, as the level of detail allows decision-makers to identify potential hazards and mitigate them
- **Sustainability** that can be conceptualised and planned for, as VDC can support assessment of energy efficiency, carbon emissions, environmental impact, embodied carbon, etc. It can not only help incorporate strategies like, adaptive reuse of materials, but also aid design of improved IEQ and plan for future use.

The VDC is a combination of several tools and can to be tailored, such seen in the Singapore VDC Framework below which boasts the slogan "*Build Twice*" (*first Virtual, then real*) ;

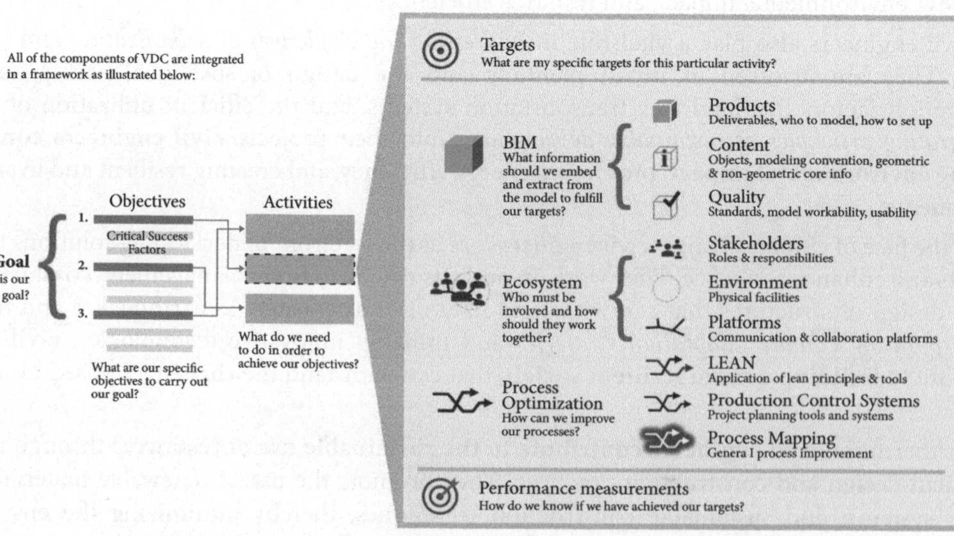

Fig. 6.6: Singapore VDC Framework 2017 (www.corenet.gov.sg)

In essence, this technology – which is a culmination of all the understanding, knowledge, discussions, and practical applications so far explored in this book, bringing sustainability hand in hand with design and construction, is the apt point for reflecting on the global and societal impact of civil engineering and the great onus on the future civil engineer.

UNIT SUMMARY

This concluding unit on projects in civil engineering covers the procedures and techniques directly applicable to construction and related activities during practice. The specific procedure for 'environmental clearance' that is mandated in India as per EIA Notification,2006, is outlined and Sustainable construction practices, with techniques for reducing GHG emissions and improving sustainability are covered to inform the requirements and consideration while proposing a project. Further, the elements of Project management, its paradigms, responsibilities, and overall contributions are discussed to give an overview on the systemic connections between decision-making during a project. Finally, the latest innovations for supporting projects are presented to make the aspiring civil engineer aware of the current trends in the domain.

EXERCISES

Q.1 What is an important element of an Environment Management Plan?

a. Probable source of Impact & Predicted Impact
b. Environmental components (ambient air, noise , water, land aesthetics)
c. Mitigation Methods
d. all of the above

Q.2 Which of the following are not objectives of sustainable construction?

a. Utilise renewable and recyclable materials
b. Green Building Certification
c. Minimize on-site waste generation
d. Safeguard natural habitats

Q.3 Which of the following techniques for reduction of GHG emissions requires Environmental Product Declaration information?

a. Minimising waste
b. Limiting carbon-intensive materials,
c. Choosing lower carbon alternatives
d. Reusing materials

Q.4 What is the traditional paradigm for Project management, where a linear sequence of phases is followed, called?

a. Agile paradigm
b. Lean construction paradigm
c. Design-Build paradigm
d. Waterfall paradigm

Q.5 Which Indian National body instituted 'The Building and other construction workers (Regulation of employment and conditions of service) Act, 1996'?

a. National Occupational Safety and Health (OSH)
b. The Ministry of Environment, Forest and Climate Change (MoEF)
c. Occupational Safety and Health Administration (OSHA)
d. Environmental Protection Agency (EPA)

Answers of Multiple Choice Questions				
6.1 **d**	6.2 **b**	6.3 **c**	6.4 **d**	6.5 **a**

I. Short and Long Answer Type Questions

Q.6 Why is Environmental impact Assessment and Environmental Clearance important consideration in the construction of built-environment.

Q.7 What is Project management? Illustrate the different types of construction management paradigms and opine which is most suitable.

Q.8 Discuss the various steps involved to get Environmental Clearance for a project in India. When is preparation of an EIA report not needed?

Q.9 What is Sustainable Construction? Elaborate on the various strategies and techniques to reduce GH emissions and improve overall sustainability.

Q.10 What are some of the new and innovative innovations and methodologies for sustainability incorporation into construction projects? Explain anyone that can be considered as a way of the future to ensure environmental sustainability.

KNOW MORE

Case studies for EIA

https://www.adb.org/publications/environmental-impact-assessment-developing-countries-asia

https://www.adb.org/projects/documents/bhu-50165-002-eia

Project Management books

https://pdfcoffee.com/construction-project-management-theory-and-practices-by-kumar-neeraj-jha-pdf-free.htm https://www.scribd.com/document/258603428/Construction-Project-Managamentby-KumarNeeraj-Jha

Environmental Clearance

https://environmentclearance.nic.in/writereaddata/Form-1A/HomeLinks/building-construction_may-10.pdf

National Occupational Safety and Health (OSH)

https://dgfasli.gov.in/sites/default/files/service_file/Nat-OSH-India-Draft%281%29.pdf

RICS Research – Real Estate and Construction Professionals in India by 2020

http://www.inskills.co.in/download/Sectors/RICS%20Report%20on%20Construction%20Skill%20Gap%202020.pdf

REFERENCES AND SUGGESTED READINGS

Books and Publications

- [Book] Shadan, K., & Fleming, G. (2012). *Construction Project Management Handbook*: March 2012 (No. FTA Report No. 0015). United States. Federal Transit Administration.
- [Book] Kibert, C. J. (2016). *Sustainable construction: green building design and delivery.* John Wiley & Sons.

- [Book] Levy, S. M. (2018). *Project management in construction*. McGraw-Hill Education
- Bugalia, N., Maemura, Y., & Ozawa, K. (2019). *Safety culture in high-speed railways and the importance of top management decisions.*
- Cleland, D. I. (1977). *Defining a project management system*. Project Management Quarterly, 8(4), 37–40.
- Kibert, C. J. (2007). *The next generation of sustainable construction*. Building Research & Information, 35(6), 595-601.

Web sources

- https://www.cseindia.org/environmental-clearance—-the-process-403
- https://www.aia.org/articles/70446-ten-steps-to-reducing-embodied-carbon
- https://dgfasli.gov.in/sites/default/files/service_file/Nat-OSH-India-Draft%281%29.pdf
- https://environmentclearance.nic.in/writereaddata/Form-1A/HomeLinks/building-construction_may-10.p

REFERENCES FOR FURTHER LEARNING

- Andersson, L., Farrell, K., Moshkovich, O., & Cranbourne, C. (2016). *Implementing Virtual Design and Construction using BIM: Current and future practices*. Routledge.
- Benedict, M. A., & McMahon, E. T. (2012). *Green infrastructure: linking landscapes and communities*. Island press.
- Carroll, J. E. (2012). *Sustainability and spirituality*. State University of New York Press.
- Costanza-Chock, S. (2020). *Design justice: Community-led practices to build the worlds we need*. The MIT Press.
- Ching, F. D. (2023). *Architecture: Form, space, and order*. John Wiley & Sons.
- Dameri, R. P., & Rosenthal-Sabroux, C. (Eds.). (2014). *Smart city: How to create public and economic value with high technology in urban space*. Springer.
- Doran, D., & Cather, B. (Eds.). (2013). *Construction materials reference book*. Routledge.
- Dover, J. W. (2015). *Green infrastructure: incorporating plants and enhancing biodiversity in buildings and urban environments*. Routledge.
- Elliot, J.A., (2006) *An introduction to sustainable development*. Routledge.
- Forbes, L. H., & Ahmed, S. M. (2010). *Modern construction: lean project delivery and integrated practices*. CRC press.
- Grober, U., & Cunningham, R. (2012). *Sustainability: A cultural history* (p. 156). Cambridge: Green Books.
- Hardi, P., & Zdan, T. (2001). *Assessing sustainable development: principles in practice*.
- Ingram, J. (2020). *Understanding BIM: The past, present and future*. Routledge.
- Jenks, M., & Jones, C. (Eds.). (2009). *Dimensions of the sustainable city*.
- Kibert, C. J. (2016). *Sustainable construction: green building design and delivery*. John Wiley & Sons.
- Lawson, M., Ogden, R., & Goodier, C. (2014). *Design in modular construction*. CRC Press.
- Lechner, N. (2014). *Heating, cooling, lighting: Sustainable design methods for architects*. John wiley & sons.
- Smith, R. E. (2010). *Prefab architecture: A guide to modular design and construction*. John Wiley & Sons.
- Vale, L. (2014). *Architecture, power and national identity*. Routledge.
- Vezzoli, C., & Manzini, E. (2008). *Design for environmental sustainability*. London: Springer
- Wang, S. (2009). *Intelligent buildings and building automation*. Routledge.

CO AND PO ATTAINMENT TABLE

Course outcomes (COs) for this course can be mapped with the programme outcomes (POs) after the completion of the course and a correlation can be made for the attainment of POs to analyze the gap. After proper analysis of the gap in the attainment of POs necessary measures can be taken to overcome the gaps.

Table for CO and PO attainment

Course Outcomes	Attainment of Programme Outcomes (1- Weak Correlation; 2- Medium correlation; 3- Strong Correlation)											
	PO-1	PO-2	PO-3	PO-4	PO-5	PO-6	PO-7	PO-8	PO-9	PO-10	PO-11	PO-12
CO-1												
CO-2												
CO-3												
CO-4												
CO-5												
CO-6												
CO-7												

The data filled in the above table can be used for gap analysis.

INDEX